Praise for
A REDDER SHADE OF GREEN

"**More than just a series of incisive contributions** that seeks to integrate Marxist social science and Earth System science, *A Redder Shade of Green* is a much-needed call for a new scientific ecosocialism of the 21st century." —FEDERICO FUENTES, editor, *Links International Journal of Socialist Renewal*

"**Beautifully written, engaging, and illuminating**, these essays offer a strong case for ecosocialism as a fusion of the sciences of nature and an updated Marxism, both recast now under the shadow of the Anthropocene."—JOHN FORAN, professor of sociology and environmental studies, University of California, Santa Barbara

"**Ian Angus has long been at the forefront** of bringing together the socialist and environmental movements, and these thought-provoking essays demonstrate his wonderful ability to make complex scientific and political ideas accessible. For those fighting for a sustainable society in the face of fossil fuel capitalism, they offer important insights, powerful polemics, and much food for further debate."—MARTIN EMPSON, author of *Land and Labour: Marxism, Ecology and Human History*

A REDDER SHADE
OF GREEN

Intersections of Science and Socialism

IAN ANGUS

MONTHLY REVIEW PRESS
New York

Copyright © 2017 by Ian Angus
All Rights Reserved

Library of Congress Cataloging-in-Publication Data
available from the publisher

isbn (paper): 978-1-58367-644-8
isbn (cloth): 978-1-58367-645-5

Typeset in Dual, and Minion Pro 11/14

Monthly Review Press, New York
monthlyreview.org

5 4 3 2 1

Contents

To Evelyn Reed (1905–1979) and George Novack (1905–1992),
activists and educators who devoted their lives
to science and socialism.

Essential Debates at the Intersections of Science and Socialism

One basis for life and another for science is a priori a lie.
—KARL MARX[1]

Red is the color of socialist revolution. It stands for egalitarian democracy and human liberation, for an end to capitalism and to all forms of oppression.

Green is the color of ecological revolution. It stands for global sustainability and a world where humans live in harmony with the rest of nature.

Red and green together are the colors of ecosocialism, a body of ideas and a movement whose fundamental principle is that there can be no true ecological revolution that is not socialist, and no true socialist revolution that is not ecological.[2]

Ecosocialism aims to build red-green movements based on a social-ecological understanding of the global crisis, to inform and advance radical movements for social and ecological justice. *A Redder Shade of Green* is a contribution to that effort.

THE ESSAYS, INTERVIEWS, AND TALKS collected here discuss some of the intersections of socialism and science that define ecosocialist politics in the twenty-first century. They cover a variety of topics, united by the conviction that ecosocialism cannot succeed unless it is more than a pious wish for a better world: it must be based on a concrete scientific understanding of how our world has evolved, how it is changing today, and where those changes may take us. The way we build socialism, and the kind of socialism that can be built, will be profoundly shaped by the state of the planet we must build it on. If our political analysis and program doesn't have a firm basis in the natural sciences, our efforts to change the world will be in vain.

The reverse is also true, because the natural sciences only reveal parts of reality. Chemistry and climatology show us that greenhouse gases are rapidly disrupting climate patterns that have existed longer than human beings have walked the earth, and they identify fossil fuel combustion as the primary source of those gases. But by themselves they cannot show why the most powerful governments and corporations have actively resisted any serious action to reduce emissions, nor can they show us how to overcome that resistance. An understanding of Earth System science is *necessary* for preventing environmental crises, but it is not *sufficient*.

For Karl Marx and Frederick Engels, socialism had to be based on *Wissenschaft*—a word that is usually translated as *science*, but which has a broader meaning in German: the systematic pursuit of all kinds of knowledge. Marx and Engels used the term "scientific socialism" not to suggest that it was comparable to chemistry or physics, but as a contrast to the utopian socialisms of the early nineteenth century, which were based on abstract moralism, not on systematic study of capitalism and its material context. For them, there was no wall between social and natural science.

Unfortunately, most of what passes for social science in universities today has abandoned any thought of a unified understanding of human society and the rest of the natural world. Even within environmental sociology, where the best scholars retain that objective, there are many who view human society as somehow exempt from the laws that govern the rest of the Earth System.

In contrast, ecosocialism seeks and is based on a careful and deliberate synthesis of Marxist social science and Earth System science—a twenty-first century rebirth, if you will, of scientific socialism.

A REDDER SHADE OF GREEN IS A BOOK of debates, polemics, and arguments, because although environmentalists, scientists, and socialists share concerns about the devastation of our planet, we frequently differ on explanations and solutions. We disagree not just with defenders of the status quo, but with one another, and that's healthy. In movements for change, unanimity is more often a sign of lack of thought than genuine agreement. We can only reach real consensus through frank and open debate, and by testing our ideas in actual campaigns to stop environmental destruction.

We don't argue for argument's sake: these are *essential debates*, part of the process of developing the theory, program, and practice of socialism in the twenty-first century.

One book cannot possibly address all the contentious issues that face ecosocialists today, so I have selected topics that have been of particular interest to readers of the web journal I edit, *Climate & Capitalism*. For clarity I have grouped these topics under five headings.

1. *Natural Science and the Making of Scientific Socialism* discusses how Marx and Engels responded to scientific developments in their time. Of course, the fact that they did or said something does not prove it is correct or should be emulated today, but an essential starting point for developing a modern Marxist approach to science is to understand how Marx and Engels approached it.

2. *Responding to the Anthropocene* addresses the unprecedented challenges posed by the new planetary epoch, and responds to critics on the right and left who are trying to de-legitimize Anthropocene science and convince environmentalists to ignore or reject what Earth System scientists are telling us.

3. *Numbers Are Not Enough* considers aspects of the overpopulation myth that Simon Butler and I were not able to address in our book *Too Many People?*, and discusses why populationist ideology continues to attract support from people who are sincerely committed to building a better world.

4. *Saving Species, Saving Oceans* reviews two important and very different books on how human activity is destroying the habitats of the animals with which we share the planet, and what can be done to prevent catastrophic declines in the wildlife that depends on forests and oceans.

5. *Toward an Ecological Civilization* responds to claims that environmental movements are being weakened by "catastrophism," discusses the role of ecosocialists in the fight for climate justice, and proposes a broad vision of the fight for a world in which the lives of future generations take precedence over today's corporate profits.

✂

THE ARTICLES ON WHICH THE CHAPTERS in *A Redder Shade of Green* are based were originally published between 2009 and 2017. All have been edited for clarity and consistency and to minimize repetition. Some are virtually unchanged, others have been extensively revised and updated. Each stands alone, but the arguments will be clearer if you read the chapters within each section in order.

Nothing here should be viewed as ecosocialist gospel or even as my final judgment on any of these subjects. These are contributions to critical debates that I fully expect to continue. I look forward to receiving responses, criticisms, and disagreements, as submissions to the online journal *Climate & Capitalism* and, more informally, on our Facebook page. I reserve the right to reply—*and* the right to change my mind in response to convincing arguments.[3]

PART ONE

NATURAL SCIENCE AND THE MAKING OF SCIENTIFIC SOCIALISM

From the beginning, Marxism took science
extremely seriously, not only for its economic promise in
building a socialist society, but for its revelatory
power in understanding the world.
—HELENA SHEEHAN[1]

CHAPTER 1

Marx and Engels and the Red Chemist

I n the closing decades of the twentieth century, a peculiar idea took hold in parts of academia. It contended that far from being the closest of comrades and collaborators who worked in harmony for forty years, Karl Marx and Frederick Engels in fact disagreed with each other on fundamental questions of theory and practice.

Their supposed disagreements concerned nature and the natural sciences. Paul Thomas, for example, contrasts "Engels's well-known interest in the natural science" with "Marx's lack of concern," and treats the divergence as evidence that "Marx and Engels are separated by a conceptual chasm that should have resisted all attempts at papering it over."[1] Terence Ball says, "The idea (later espoused by Engels) that nature exists independently of, and prior to, man's efforts to transform it is utterly foreign to Marx's humanism."[2] In Ball's view, Engels's distortions of Marxist philosophy were directly responsible for "some of the more repressive features of Soviet practice."[3] In a more extreme version, Terrell Carver and others insist that Marx was not a Marxist—that

---

Adapted from Ian Angus, "Marx And Engels and the Red Chemist," *Monthly Review*, March 2017.

Marxism was a doctrine invented by Engels, whose scientific materialism was contrary to Marx's liberal humanism.

From a somewhat different perspective, Theodor Adorno, Alfred Schmidt, and others associated with the Frankfurt School and Western Marxism have argued that historical materialism applies only to human society, so Engels's efforts to adapt it to the natural sciences in his unfinished work *Dialectics of Nature* were intellectual distortions that went against the Marxist method.

Engels's defenders have countered that Marx and Engels had a division of labor in which Engels studied science and Marx did not, but a growing body of research shows that that response gives too much ground to the anti-Engels argument. As Kohei Saito writes, the division of labor is an illusion: "Though Engels is better known for his writings on natural science . . . Marx was an equally keen student of many of the same subjects."[4]

New studies of Marx's long-unavailable notebooks, now being published in the massive *Marx-Engels-Gesamtausgabe* (Marx-Engels Complete Works), decisively refute claims that Marx was uninterested in the natural sciences or considered them irrelevant to his politics.

> Marx's notebooks allow us to see clearly his interests and preoccupations before and after the publication of the first volume of *Capital* in 1867, and the directions he might have taken through his intensive research into disciplines such as biology, chemistry, geology, and mineralogy, much of which he was not able fully to integrate into *Capital*. While the grand project of *Capital* would remain unfinished, in the final fifteen years of his life Marx filled an enormous number of notebooks with fragments and excerpts. In fact, a third of his notebooks date to this period, and almost one-half of them deal with natural sciences. The intensity and scope of Marx's scientific studies is astonishing.[5]

As more of this material becomes available for study, it may shed new light on the role of Carl Schorlemmer, a working scientist whose important contributions to the development of scientific socialism have been unjustly ignored. Most English-language accounts of Marx and Engels's lives, if they mention Schorlemmer at all, refer to him only as a friend, without acknowledging his influence on their studies of the natural sciences.[6] It is time to restore this neglected figure to his rightful place in the Marxian— and Engelsian—tradition.

Chemist and Comrade

Carl Schorlemmer was born in 1834 in Darmstadt, in the Rhine-Main region of what is now Germany. The son of a carpenter, he studied pharmacy at the Darmstadt technical college and chemistry at the University of Giessen. In 1859 he was hired as an assistant to chemistry professor Henry Roscoe at Owens College in Manchester, where he lived for the rest of his life.

Schorlemmer was one of the most respected and gifted chemists of his time. In his first decade in Manchester he published more than two dozen scientific papers, many of them groundbreaking studies of hydrocarbon chemistry. He was elected a Fellow of the Royal Society in 1871, and appointed to England's first chair of organic chemistry at Owens College in 1874. He served as vice president of the chemical section of the British Academy in the 1880s, and when he received an honorary doctorate from the University of Glasgow in 1888, the presenter described him as "one of the greatest living writers and authorities on organic chemistry."[7] After his death, Owens College spent £4,800—equivalent to more than $1 million today—to build and equip the Schorlemmer Memorial Laboratory, the first facility in England devoted solely to organic chemistry.

And he was a communist.

Frederick Engels lived in Manchester from 1850 to 1870, working in a textile company that was half owned by his father. He hated the work, but living in the world's first great industrial city gave him opportunities to learn about the English and Irish working class, guided by his common-law wife, Mary Burns, and about the development of industrial capitalism. It was probably the latter study that led him to the Thatched House Tavern, where young German scientists who worked in the chemical industries around Manchester met periodically to discuss science, business and industry, and, inevitably in a group of expatriates, German politics. At one of those meetings, probably in early 1865,[8] Engels met Carl Schorlemmer, whom he described to Marx as "one of the best fellows I have got to know for a long time."[9]

Schorlemmer quickly became Engels's closest friend in Manchester, and was welcomed into Marx's extended family circle in London. His sense of humor led Marx to dub him Jollymeier, a nickname that stuck for the rest of his life. He was a frequent visitor in the Marx and Engels family homes, and spent most of his summer vacations with them in London or at the seaside. After Marx's death in 1883, Engels and Schorlemmer spent even more time together: they spent two months travelling in the United States and Canada in 1888, and several weeks in Norway in 1890. The chemist's death in 1892 was, as Eleanor Marx wrote to her sister Laura, "a very great blow" to Engels.[10]

In his account of the gathering at Marx's graveside in 1883, Engels wrote, "The natural sciences were represented by two celebrities of the first rank, the zoology Professor Ray Lankester and the chemistry Professor Schorlemmer, both members of the Royal Society in London."[11] In a subsequent letter to Eduard Bernstein, Engels described Schorlemmer as "after Marx, undoubtedly the most eminent man in the European socialist party."[12]

Marx, Engels, and Schorlemmer were thus more than friends—they shared a common political commitment and social vision. Engels later recalled that when they met, the chemist was already "a complete communist," and had been for some years: "All he had to learn from us were the economic grounds for a conviction he had gained long ago."[13] As part of the teaching process, Engels shared with him the proof sheets of volume one of *Capital*, before Marx's masterpiece was published.

Schorlemmer joined the International Workingmen's Association (the First International) and the German Social Democratic Party. He allowed his home address to be used for correspondence and parcels when it appeared that the police were opening letters between Marx and Engels, and during trips to scientific conferences on the continent he helped strengthen links with socialists there. On one such visit in 1883, the police detained him and searched his family's home because his presence at Marx's funeral had been mentioned in the socialist press. He was suspected of (and probably was) smuggling banned socialist literature into Germany.

But Schorlemmer's most lasting contribution was helping Marx and Engels to understand the latest developments in the natural sciences. "A chemist has recently explained Tyndall's experiment with light to me," was Engels's first mention of his new friend to Marx.[14] Many more explanations would follow.

Unfortunately, their relationship is mostly undocumented. For the first five years of their friendship, Schorlemmer and Engels both lived in Manchester, so there were no letters between them. After Engels moved to London in 1870, "the greater part of our lively correspondence was concerned with the sciences and party affairs," but none of those letters have survived.[15] As a result, we must depend on letters between Marx and Engels for a valuable, if incomplete, picture of their long collaboration with the man who has been called the "Red Chemist."[16]

Dialectics and Science

Marx's research for *Capital* had included careful study of Justus von Liebig's work on agricultural chemistry, which he described as "more important for this matter than all the economists put together."[17] John Bellamy Foster has shown that this research was central to the development of Marx's concept of a "metabolic rift" between capitalist society and nature.[18] Shortly after volume one of *Capital* was published in 1867, Marx asked Schorlemmer, through

Engels, for help in continuing those studies. Having "taken an extraordinary liking to Schorlemmer's compendium"—a reference to Schorlemmer's German translation and revision of Roscoe's *Short Textbook on Chemistry*[19]—Marx wrote to Engels:

> I would like to know from Schorlemmer what is the latest and best book (German) on agricultural chemistry. Furthermore, what is the present state of the argument between the mineral-fertiliser people and the nitrogen-fertiliser people? (Since I last looked into the subject, all sorts of new things have appeared in Germany.) Does he know anything about the most recent Germans who have written *against* Liebig's soil-exhaustion theory? Does he know about the alluvion theory of the Munich agronomist Fraas (Professor at Munich University)? For the chapter on ground rent I shall have to be aware of the latest state of the question at least *to some extent*. Since Schorlemmer is an expert on this subject he must be able to give information.[22]

No one could reasonably describe the author of that letter as exhibiting "lack of concern" about science!

This particular request for scientific information didn't pay off. Schorlemmer's response contained little that was new to Marx, but it was a start.[21] Marx met Schorlemmer in 1868 when the chemist visited London to present a paper on organic chemistry to the Royal Society, and thereafter addressed his queries directly. In 1870, for example, he asked Schorlemmer's opinion about an article in the newspaper *La Marseillaise* regarding the possibility of making gun-cotton for military and mining purposes.[22]

Schorlemmer wasn't just a source of facts. Marx and Engels quickly learned that their friend's interests and expertise went well beyond his pioneering research on hydrocarbons. As Engels wrote in a memorial tribute:

> Apart from this speciality of his . . . he also devoted a great deal of attention to what is called theoretical chemistry, i.e., to

the basic laws of this science, and the way it fits in with related sciences, that is to say physics and physiology. He was particularly capable in this field. He was probably the only important scientist of his time who did not disdain learning from Hegel.[23]

We can see that side of him in May 1873, when Engels wrote to Marx, who was visiting Manchester, asking for comment on some "dialectical points about the natural sciences" that had occurred to him:

> The object of science: matter in motion, bodies. Bodies cannot be separated from motion, their forms and kinds can only be apprehended in motion; nothing can be said about bodies divorced from motion, divorced from all relation to other bodies. Only in motion does the body reveal what it is. Hence natural science obtains knowledge about bodies by examining them in their relationship to each other, in motion. Cognition of the various forms of motion is cognition of bodies. The investigation of these various forms of motion is therefore the chief object of the natural sciences.
>
> 1. The simplest form of motion is change of *place* (within time, to do old Hegel a favour)—*mechanical* motion.
>
> a) There is no such thing as the motion of a *single* body, but relatively speaking one can talk of a *falling* body in such terms. Motion towards a centre common to many bodies. However, as soon as a single body moves in a direction *other* than towards the centre, the laws of *falling* still apply, it is true, but they are modified
>
> b) into the laws of trajectories and lead directly to the interaction of several bodies—planetary, etc., motion, astronomy, equilibrium—temporarily or apparently to motion itself. The *real* result of this kind of motion, however, is ultimately always the contact of the moving bodies; they fall into one another.
>
> c) Mechanics of contact—bodies in contact. Ordinary mechanics, levers, inclined planes, etc. But *this does not exhaust*

the effects of contact. Contact is manifested directly in two forms: friction and impact. It is a characteristic of both that, at certain degrees of intensity and under certain conditions, they give rise to *new* effects, no longer of a merely mechanical nature: *heat, light, electricity, magnetism.*

2. *Physics proper*, the science of these forms of motion, establishes the fact, after investigating each form of motion separately, that they *merge into one another* under certain conditions, and ultimately discovers that given a certain degree of intensity, which varies for different moving bodies, they all produce effects that transcend physics; changes in the internal structure of bodies—*chemical* effects.

3. *Chemistry*. In the case of the preceding forms of motion it was more or less immaterial whether the bodies under investigation were animate or inanimate. In fact inanimate objects revealed the nature of the phenomena involved in their greatest *purity*. Chemistry, on the other hand, can only discover the chemical nature of the most important bodies by using substances deriving from the life process. Its principal task is increasingly that of producing these substances synthetically. It forms the transition to the science of organisms, but the dialectical point of transition can only be constructed when chemistry has effected the real transition or is on the point of doing so.

4. Organism—for the present I shall not indulge in any dialectical speculations on the subject.[24]

Engels concluded that "since you are at the very centre of the natural sciences"—a jocular reference to the fact that Marx was staying in the rooming house where Schorlemmer lived—he would be in the best position to judge whether these thoughts made sense.

These were indeed *dialectical* points. All matter is in constant motion. Under certain conditions, change in place or time becomes qualitative change—mechanical energy transforms into heat, light, electricity, or magnetism. Physical changes in turn produce chemical changes, and chemical changes give rise to living organisms. At each "dialectical point of transition," something entirely new emerges.

If Marx actually rejected applying dialectical analysis to non-human nature, as some critics claim, Engels would likely have expected and received sharp criticism, for Marx was never reluctant to argue. But Marx didn't object, nor did he assume that his own understanding of dialectics qualified him to judge Engels's thoughts on natural science. Instead he replied that he had learned a lot from Engels's letter—it "edified me greatly"—but would "venture no judgment until I have had time to reflect on the matter and consult the 'authorities.'" The "authorities," of course, meant Carl Schorlemmer. His letter concludes, "Schorlemmer read your letter and says that he is essentially in agreement with you but reserves his judgment on points of detail."[25]

He returned Engels's letter with marginal notes added by the chemist. Beside point number 1, about the investigation of matter in motion being the chief object of the natural sciences, Schorlemmer wrote, "Very good; my own view. C.S." Beside the discussion of mechanical motion, "Quite right." Beside the paragraph on chemical change leading to life, "That's the point!" And beside Engels's comment that he would not speculate on organisms for now, Schorlemmer lived up to his Jollymeier nickname by scribbling "Nor shall I. C.S." [26]

The three communists discussed these and related topics at length during their frequent visits to each other's homes. Engels's "dialectical points," in more developed and detailed form, are key themes in his two major projects of the 1870s—the incomplete drafts and notes that were published long after his death as *Dialectics of Nature* and the very influential *Herr Eugen Dühring's Revolution in Science* (better known as *Anti-Dühring*), written and published in 1876–78, which deals extensively with dialectics and natural science. Although Schorlemmer's contributions to those works were not acknowledged in the text—possibly out of concern that mention in a communist book might harm his career or endanger his family in Germany—they were undoubtedly substantial.

Engels later wrote that he read *Anti-Dühring* aloud to Marx before it was published and Marx agreed with it. Critics who cite *Anti-Dühring* as an example of Marx and Engels's disagreements have accused Engels of lying about that. Paul Thomas, for example, calls it "a curious claim, since Marx was not incapacitated or bedridden at the time it was written, and listening to a recitation of its ponderous contents would have taxed the patience of Job."[27] Bear in mind, however, that *Anti-Dühring* was written and published in installments over two years, and given what we know of their working relationship, it is likely that Engels and Marx, and Schorlemmer when he was in London, read and discussed the chapters together. The easiest way to do that—at a time when there were no typewriters or copiers—would have been for Engels to begin by reading his handwritten drafts aloud.

Note also that Marx himself wrote the chapter on economics in *Anti-Dühring*, and that in 1880 he wrote the introduction to three chapters of the book that were published separately as *Socialism: Utopian and Scientific*. He would scarcely have done either for a book he had not read or with which he fundamentally disagreed.

This years-long intellectual exchange among Marx, Engels, and Schorlemmer was not one-way. Marx's copy of Schorlemmer's *Treatise on Chemistry* includes a handwritten note from the author on the title page, acknowledging "many corrections and some suggestions . . . by the hand of Karl Marx."[28]

In 1885 Schorlemmer told a reporter from the journal *Revue Scientific* that "if chemists ever succeed in artificially obtaining proteins, they will be in the form of living protoplasm" and "the enigma of life can only be resolved by protein synthesis."[29] That concept was borrowed directly from *Anti-Dühring*. Engels commented that by presenting that speculation as his own view, "Schorlemmer did a bold thing, for if it falls flat, the blame will be his, whereas if it catches on, he will be the first to give me the credit."[30]

Later, when preparing the second edition of his book *The Rise and Development of Organic Chemistry,* Schorlemmer added a quotation from *Anti-Dühring*—giving Engels full credit—to explain how quantity changes to quality in the various hydrocarbons: "Each new member is brought into existence by the addition of CH_2 to the molecule of the preceding one, and this quantitative change in the molecule produces every time a qualitatively different body."[31]

After Schorlemmer's death in 1892, Engels wrote that it would be difficult to find a biographer who could do justice to his life. It would require someone "who is not only a chemist but also a Social Democrat, and not only a Social Democrat but also a chemist and, what's more, a chemist who has made a close study of the history of his discipline since Liebig's day."[32] That, of course, was a description of Schorlemmer himself, a man whose unique talents allowed him to work with Marx and Engels in collaboratively developing a body of thought that seamlessly integrated insights from the social and natural sciences.

Bridging the Gulf

In a famous lecture delivered at Cambridge University in 1959, British novelist C. P. Snow decried the division of the academic world into "two cultures," with the natural sciences on one side and the humanities on the other, divided by a "gulf of mutual incomprehension." Though he criticized the narrowness of both sides, he was particularly outraged by the arrogant dismissal that scholars in the humanities displayed toward science and scientists.

At the time Snow spoke, parts of the academic left in Western Europe and North America were also infected by a version of "two cultures" virus, and the disease has spread and worsened since. When a prominent left-wing academic calls ecology "the new opiate of the masses," and another describes the Anthropocene as "the most dangerous environmentalist concept of our times"[33]—and neither appears to have read the relevant scientific literature—then we are clearly dealing with anti-science prejudice, not reasoned analysis.

The relationship with Carl Schorlemmer shows that Karl Marx and Frederick Engels would have had no patience with attempts to build political or philosophical walls between the social and natural sciences. As they wrote in 1846, in the first fully developed account of historical materialism:

> We know only a single science, the study of history. One can look at history from two sides and divide it into the history of nature and the history of men. The two sides are, however, inseparable; the history of nature and the history of men are dependent on each other so long as men exist.[34]

When Marx and Engels wrote that, they were developing the philosophical basis for scientific socialism. A defining principle of that philosophy was that thought isolated from practice is meaningless: "Man must prove the truth, i.e., the reality and power, the this-worldliness of his thinking in practice."[35] What they were really setting out might best be viewed as a set of *hypotheses* that required testing in the real world.

For the rest of their lives, therefore, they tested their ideas about how the world works and changes through active participation in workers' movements and through intense study of history, economics, and the natural sciences. They did this not merely to satisfy their intellectual curiosity, though both men had an abundance of that, but because they knew it was impossible to understand and combat capitalism without a thorough understanding of the material context within which it had developed and could change in the future.

For them, and for Carl Schorlemmer, the scientific study of nature was inseparable from politics, economics, and the struggle for a better world.

CHAPTER 2

Marx and Engels and Darwin

Only 1,250 copies of the first edition of Charles Darwin's *On the Origin of Species* were printed in 1859, and they all sold in one day. One person who obtained a copy was Frederick Engels, then living in Manchester. Three weeks later, he wrote to Karl Marx:

> Darwin, by the way, whom I'm reading just now, is absolutely splendid. There was one aspect of teleology that had yet to be demolished, and that has now been done. Never before has so grandiose an attempt been made to demonstrate historical evolution in Nature, and certainly never to such good effect.[1]

When Marx read *Origin* a year later, he was just as enthusiastic, calling it "the book which contains the basis in natural history for our view."[2] In a letter to the German socialist Ferdinand Lassalle, he wrote:

Adapted from Ian Angus, "Marx and Engels . . . and Darwin?" *International Socialist Review*, May–June 2009.

Darwin's work is most important and suits my purpose in that it provides a basis in natural science for the historical class struggle. . . . Despite all shortcomings, it is here that, for the first time, "teleology" in natural science is not only dealt a mortal blow but its rational meaning is empirically explained.[3]

In 1862 Marx made a point of attending the public lectures on evolution given by Darwin's supporter Thomas Huxley, and encouraged his political associates to join him. Wilhelm Liebknecht, a friend and comrade who often visited the Marx family in London, later recalled, "When Darwin drew the conclusions from his research work and brought them to the knowledge of the public, we spoke of nothing else for months but Darwin and the enormous significance of his scientific discoveries."[4]

Although Marx and Engels criticized various aspects of Darwin's "clumsy English style of argument," they retained the highest regard for his scientific work for the rest of their lives. In his own masterwork, Marx described *On the Origin of Species* as an "epoch-making work."[5] In 1872 Marx sent a copy of *Capital* to Darwin, inscribing it "on the part of his sincere admirer, Karl Marx."[6]

And in 1883, at Marx's funeral, Engels said, "Just as Darwin discovered the law of development of organic nature, so Marx discovered the law of development of human history."[7]

Stealing the Darwinian Mantle?

Charles Darwin, once condemned as a dangerous atheist, is today the object not only of veneration but of a "Darwin industry" of academics and others who churn out an endless stream of books and articles about every possible aspect of his life and work.

But Marx and Engels are still beyond the pale, so it's not surprising that some in the Darwin industry argue that there is no real connection between Darwinism and Marxism. Marx and Engels, the claim goes, were illegitimately trying to hitch their wagon to Darwin's star. Among others,

- Allan Megill argues that "Marx and Engels were willing to appeal to Darwinism for propaganda purposes," but any impression that Darwinian evolution is similar to Marxism is "totally false."[8]
- Naomi Beck claims that for Marx and Engels, "Darwin's theory fulfilled for them only the function of a pretext and was not in reality connected with their views." Engels's comparison of Marx and Darwin was just an opportunist attempt to "establish Marx's independent scientific status as Darwin's equal."[9]
- D. A. Stack says that Engels's remarks at Marx's graveside were part of a "parochial propagandist campaign to steal the Darwinian mantle. . . . The term 'Darwinian' was sought as an honorific title, nothing more." Engels was just "keen for Marxism to bask in the reflected glory of Darwinism."[10]

It's difficult to decide which is worse, the cynicism that suggests Engels would use his lifelong comrade's funeral as an occasion to win petty political advantage or the ignorance these writers display of both the revolutionary implications of Darwinism and the importance of natural science to Marxist theory.

Anyone who seriously studies the works of Marx, Engels, and Darwin will understand, even if they don't agree with him, that Marx was both honest and exceptionally insightful when he wrote that *On the Origin of Species* "contains the basis in natural history for our view." To understand what Marx meant, we need to understand what Darwin wrote, and why his views marked a radical break with the dominant ideas of his day.

An Unlikely Revolutionary

Charles Robert Darwin was, to say the least, an unlikely revolutionary. His father was a prominent physician and wealthy investor; his grandfather was Josiah Wedgwood, founder of one of the largest manufacturing companies in Europe. He could have lived a life of leisure, but instead he devoted his life to science.

In 1825, Charles's father sent him to the University of Edinburgh to study medicine, but he was much more interested in studying nature, a subject not offered as a degree program at any university in Great Britain. After two years he dropped out of Edinburgh and enrolled in Cambridge, aiming to become an Anglican priest, a respectable profession that would allow him leisure time to collect beetles, stuff birds, or search for fossils. This wasn't as improbable as it seems today. At the time, the great majority of naturalists in England, including all the professors who taught science at Oxford and Cambridge, were ordained Anglican priests. Clergymen studied nature not just for its own sake but as a contribution to "natural theology"—understanding God by studying God's works.

Darwin seems to have been a competent theology student, but he particularly impressed the men who taught science. After graduation in 1831, one professor took him on a three-week geology expedition in North Wales, and then his botany professor recommended him to Captain Robert FitzRoy of the Royal Navy, who was looking for a gentleman naturalist to travel with him as an unpaid companion on a surveying voyage to South America and the South Pacific.[11]

And so it began. On December 27, 1831, twenty-two-year-old Charles Darwin boarded the British survey ship HMS *Beagle*. Although plagued by seasickness, he traveled much more comfortably than the crew: he ate at the captain's table, was accompanied by a manservant, and had more than sufficient funds (provided by his doting father) to rent comfortable accommodations when the ship was in port. But it wasn't a pleasure trip; he conducted extensive and detailed geological studies, wrote thousands of pages of scientific observations, and collected more than 1,500 specimens of living and fossil life.

When he left England, Darwin seems to have been a conventional Christian who agreed with "the great majority of naturalists [who] believed that species were immutable productions, and had been separately created."[12] Biblical literalists and deists alike agreed that species were fixed by divine law. Dogs might vary in appearance, but dogs don't turn into pigs or give birth to cats.

After five years of scientific research on the *Beagle* and two more years of study at home, Darwin came to a heretical conclusion: species were not immutable. All animals were descended from common ancestors, different species resulted from gradual changes over millions of years, and God had nothing to do with it.

It is difficult, today, to appreciate just how shocking this idea would be to the middle and upper classes of Darwin's time. Religion wasn't just the "opium of the masses"—it gave the wealthy moral justification for their privileged lives in a world of constant change and gross inequality.

One of the most popular hymns of the Victorian age clearly expressed the link between God the creator of all life and God the preserver of social stability:

> All things bright and beautiful,
> All creatures great and small,
> All things wise and wonderful,
> The Lord God made them all.
>
> Each little flower that opens,
> Each little bird that sings,
> He made their glowing colors,
> He made their tiny wings.
>
> The rich man in his castle,
> The poor man at his gate,
> God made them, high or lowly,
> And order'd their estate.

"All Things Bright and Beautiful" was published in 1848, shortly after a famine that killed more than a million people in Ireland, and while revolutionary uprisings were sweeping across Europe. In the face of such social crises, hymns like this and the sermons that accompanied them taught both rich and poor that the status quo was divinely ordained. Anyone who questioned God's word was endangering a very fragile social order.

The Mystery of Mysteries

Nevertheless, by the 1830s educated people, including Darwin, knew that the Genesis creation story was not literally true. The expansion of capitalism had led to booms in mining and canal building; those works exposed geological layers and ancient fossils that proved the earth was millions of years old, not the six thousand years allowed by biblical chronology. What's more, the fossil record showed that animals unknown today were once common, and that modern animals had appeared relatively recently, contradicting the claim that God created all species at once.

In the same period, imperialist expansion led to global exploration and the discovery of more varieties of plant and animal life than any European had ever imagined, far more than could have lived in Eden or found space on Noah's Ark.

By the 1830s, scientists agreed that there were only two possible explanations for the accumulating evidence. The very influential Cambridge professor William Whewell summed up the choices:

> Either we must accept the doctrine of the transmutation of species, and must suppose that the organized species of one geological epoch were transmuted into those of another by some long-continued agency of natural causes; or else we must believe in many successive acts of creation and extinction of species, out of the common course of nature; acts which, therefore, we may properly call miraculous.

Whewell, like every other respectable scientist of the time, had no doubt about the answer: animals and plants may vary in response to external circumstances, but "the extreme limit of variation may usually be reached in a brief period of time: in short, species have a real existence in nature, and a transition from one to another does not exist." [13]

If species could not change over time, only miracles could explain the fossil record. But how did God do it? What did the process of divine creation actually look like on earth? "The replacement of extinct species by others," was, wrote astronomer John Herschel, the "mystery of mysteries." [14]

While some scientists and theologians insisted that God must personally intervene each time a new species is required, others were confident that the Creator had set up the universe so that new species were created through "secondary causes"—that is, by natural means—whenever they were needed.

What's particularly noteworthy today is that "God did it" wasn't just an acceptable answer to difficult questions, it was standard scientific methodology. Even scientists who believed that nature could be completely explained by natural laws believed that God established those laws to ensure that creation proceeded according to his will.

Evolution before Darwin

The very fact that the scientific establishment thought it necessary to vigorously deny "transmutation of species" shows that not everyone agreed that species couldn't change.

A noteworthy example was Charles Darwin's grandfather, Erasmus Darwin, who described something like evolution in his 1794 book *Zoonomia*, and again in 1803 in a book-length poem, *The Temple of Nature*. His evolutionary ideas don't seem to have influenced anyone, probably the result, as Charles Darwin later wrote, of "the proportion of speculation being so large to the facts given." [15]

Others offered similar speculations, but before Darwin, only two writers proposed worked-out theories of species change over time: Jean-Baptiste Lamarck and Robert Chambers.

Lamarck was appointed head of the invertebrate division of the Musée National d'Histoire Naturelle in Paris when France's

revolutionary government reorganized the country's scientific institutions in the 1790s. In the early 1800s, he argued that all modern
animals are the descendants of less complex ancestors. Unlike
Darwin, Lamarck didn't suggest common descent, but rather a
complex model in which every type of organism went through
a separate evolutionary process. Nature, he believed, constantly
and spontaneously creates new evolutionary lines, beginning with
single-celled animals that have an innate drive to become more
complex, or perfect, over time. Eventually, if the climb isn't interrupted, they reach the peak of perfection as human beings.

But the climb is often interrupted by environmental changes
to which the animal must respond. Giraffes develop long necks by
stretching to reach high leaves, and fish that live in caves become
blind because they don't use their eyes. These changes are then
inherited by their offspring. In Lamarck's works, this was a secondary process, but the term "Lamarckism" has since come to
mean "inheritance of acquired characteristics" and nothing else.

Lamarck's views won little support from other scientists, even
in France, but there was a significant underground Lamarckian
current in England among radical democrats, socialists, and secu-
larists between 1820 and 1850. Many of them used Lamarckian
arguments to criticize the undemocratic English state and the
Anglican Church:

> Cannibalized fragments of Lamarck's evolutionary biology—
> which provided a model of relentless ascent power-driven "from
> below"—turned up in the pauper press. Lamarck's notion that
> an animal could, through its own exertions, transform itself
> into a higher being and pass on its gains—all without the aid
> of a deity—appealed to the insurrectionary working classes. His
> ideas were propagated in their illegal penny prints, where they
> mixed with demands for democracy and attacks on the clergy.[16]

Much more influential on broad public opinion in England was
Vestiges of the Natural History of Creation, published anonymously

in 1844 by Robert Chambers, a magazine publisher and amateur geologist from Edinburgh. He attributed the entire history of the universe to a God-ordained "Law of Development" that produced stars, planets, and eventually life. After the first life arose spontaneously on Earth, animals and plants ascended the ladder of life. "It has pleased Providence to arrange that one species should give birth to another, until the second highest gave birth to man, who is the very highest."[17]

Chambers meant "gave birth" literally. Drawing on the theory that embryos pass through stages that repeated the development of their ancestors, he concluded that when it was time for a new species to arrive, females would somehow extend their gestation periods, so that their offspring would emerge as the next species up the ladder.

Universally condemned by the scientific establishment at the time, and nearly forgotten today, *Vestiges* was nevertheless a sensational bestseller. Before *On the Origin of Species*, *Vestiges* was the only book on evolution that most English readers might have read.[18]

Essentialism and Teleology

As we've seen, the scientific discoveries of the late eighteenth and early nineteenth centuries provoked widespread speculation about Herschel's "mystery of mysteries." Most professional scientists and many amateurs and outsiders offered views on how the apparent extinction and creation of species could be explained or explained away. Though the explanations varied, they all rested on a common ideology, the twin concepts of essentialism and teleology.

Essentialism is based on the first law of formal logic: that a thing is always equal to itself, that A always equals A. That's a useful, even necessary assumption for many purposes, but it ignores the reality of change—that over time all things decay, or transform, or combine, so that A turns into something that is no longer A.

In nineteenth-century natural science, essentialist thinkers assumed that the definition or idea of a species is more important, indeed *more real*, than the specific organisms we can actually observe. A species is a constant, unchanging type; the variations we observe in nature are accidental and transitory.

As we've seen, William Whewell believed firmly that "species have a real existence in nature, and a transition from one to another does not exist." Charles Lyell, the leading geologist of the time, devoted several chapters of his influential book *Principles of Geology* to a critique of Lamarck and the very idea that species can change. As Stephen Jay Gould points out, Lyell's argument was rooted not in actual study of nature, but in his essentialist philosophy:

> The focus of Lyell's argument—and the reason for lambasting evolution defined as insensible transition between species—rests upon a view of species as entities, not tendencies; things, not arbitrary segments of a flux. Species arise at particular times in particular regions. They are, if you will, particles with a definite point of origin, an unchanging character during their geological duration, and a clear moment of extinction.[19]

It is obvious that those who rejected evolution held essentialist views. But people like Chambers, who held that one kind of organism could give birth to another, were also essentialist. In their view of evolution, species didn't change; rather, one natural kind was wholly replaced by a new one.

Teleology is the belief that all things are designed for or inherently directed toward a final result. Birds were given wings so that they could fly, giraffes got long necks so that they could reach high leaves, and the earth was created as a place for people to live.

The idea that Earth and everything in it was designed by God to achieve his divine ends was almost universally accepted by the leading philosophers and scientists in the nineteenth century. Serious thinkers claimed that coal deposits were laid down in England so that they could later be used by industry, or that the fact that divine design caused the life cycle of most plants to equal the duration of the earth's revolution around the sun.

Even Lamarck, who did not include God in his theory, held that there was a mysterious force driving all organisms to become ever more perfect, until they reached perfection as human beings.

Natural Selection

In *Origin*, Darwin argued that three factors combine to create new species: population pressure, variation and inheritance, and natural selection.

1. *Population pressure*: All organisms tend to have more offspring than can survive in the local environment. Many individuals either do not survive or are not able to reproduce.
2. *Variations and heritability*: There are many variations between the members of a given population: no two individuals are exactly alike. Most of these variations are inheritable—that is, they are passed on to the offspring of the individuals concerned. While most of these variations are insignificant (eye color, for example), some will increase or decrease the individual's chances of surviving and reproducing.
3. *Natural selection*: Individuals with favorable variations will tend to have more offspring than average; those with unfavorable variations will tend to have fewer. As a result, over long periods of time unfavorable variations will tend to decrease in frequency, while favorable variations will become more common.

This implied a very different explanation of the giraffe's long neck. Contrary to Lamarck, Darwin suggested that the giraffe's ancestors had necks of various lengths. Those with longer necks could reach more leaves than those with shorter necks. Being better fed, they were stronger, and tended to live longer and have more offspring— so over time the population's average neck length increased.

Unlike Lamarck and Chambers, Darwin wasn't just speculating. His "theory of descent with modification through natural selection"[20] was developed and then fine-tuned in years of careful study and experimentation. In his home in rural Kent, south of London, he dissected all kinds of animals, raised pigeons to learn about variation and inheritance, and experimented with plant germination and seed dispersal. Above all, he sought out and learned from people with practical, hands-on knowledge—gamekeepers, pigeon enthusiasts, sheep and cattle breeders, gardeners, and zoo managers.

These materialist methods led him to an entirely materialist theory, at a time when materialism wasn't just unpopular in respectable circles but was considered subversive and politically dangerous. Between 1838 and 1848, while Darwin was first working out his ideas, England was swept by an unprecedented wave of mass actions, political protests, and strikes. Radical ideas—materialist, atheistic ideas—were infecting the working class, leading many to expect (or fear) revolutionary change.

Darwin was never actively involved in politics, but he was a privileged member of the wealthy middle class and that class was under attack. As John Bellamy Foster writes, "Darwin was a strong believer in the bourgeois order. His science was revolutionary, but Darwin the man was not."[21]

Rather than risk being identified with the radicals and perhaps being ostracized by his fellow gentleman-scientists, Darwin wrote a 270-page account of his theory in 1844, attached a letter asking his wife to publish it if he died, and told no one else. Between 1840 and 1854 he wrote a popular account of his voyage around the world, scientific books on coral reefs and volcanic islands, and an exhaustive four-volume study of barnacles. Only in the mid-1850s, when his scientific reputation was assured, and the social turbulence of the 1840s was clearly over, did he return to the subject for which he is now most famous.

Even then he would likely have delayed into the next decade had not a younger naturalist, Alfred Russel Wallace, sent him an essay containing ideas similar to his own. Pressed by friends to publish, Darwin set aside "the big book on species" he was working on and prepared what he called an abstract. *On the Origin of Species by Means of Natural Selection, or, the Preservation of Favoured Races in the Struggle for Life* was published in November 1859.

Turning Science Right Side Up

Marx wrote that in Hegel's writings, the dialectic was "standing on its head," so it had to be turned right side up to discover "the rational kernel within the mystical shell."[22] That is what Marx and Engels did in the process of working out the fundamental basis of their views, historical materialism.

And that is exactly what Darwin did in *On the Origin of Species*. He overturned the fundamental concepts of nineteenth-century science, turning an upside-down view of nature right side up.

He overturned essentialism. "I look at the term 'species' as one arbitrarily given, for the sake of convenience, to a set of individuals closely resembling each other."[23]

A species is not a thing, and change does not involve the transformation or replacement of that thing. A species is a population of real, concrete individuals. Variations are not exceptions or diversions from the species' essence—variation is the concrete reality of nature. The truth, a Marxist would say, is always concrete. Species are not fixed, immutable things: they have a real history, and can only be properly understood by studying how they change in time.

He overturned teleology. "Far from imagining that cats exist in order to catch mice well," wrote Darwin's close associate Thomas Huxley, "Darwinism supposes that cats exist because they catch mice well—mousing being not the aim, but the condition of their existence."[24]

Living organisms have changed and continue to change as a result of natural processes that have no purpose or goal. A giraffe is not in any sense a "more advanced" or "more perfect" animal than its shorter ancestors—it is simply better adapted to its local environment. Changes to that environment could eliminate its advantage at any time.

By the time Darwin died in 1882, evolution was accepted by the great majority of scientists—but it took much longer for most to accept the materialist core of Darwin's work, that variation and natural selection are the processes that drive species change. Even among Darwin's closest allies and supporters there were many who clung to the essentialist idea that new species must appear as sudden replacements, or to the teleological idea that the evolutionary process was guided or determined in advance by God.

Evolution and Marxism

Darwin did for the understanding of nature what Marx and Engels did for human society. He overturned teleology and essentialism and established a materialist basis for understanding how organisms change over time. And that is precisely what Marx meant when he said that On the Origin of Species "contains the basis in natural history for our view."

In 1844, while Darwin was secretly writing his first full account of natural selection, Karl Marx was in Paris, developing his critique of contemporary political and philosophical thought. In his notebooks he wrote:

> History itself is a real part of natural history and of nature's development into man. Natural science will, in time, incorporate into itself the science of man, just as the science of man will incorporate into itself natural science: there will be one science.[25]

Such comments show why Marx and Engels were so excited by Darwin's work. As Paul Heyer writes, "Both the historical character of nature and the natural character of history" were fundamental to their worldview.[26]

Fifteen years before Origin, Marx and Engels were confident that nature could be explained using the same non-essentialist and non-teleological—that is, historical and materialist—principles that underlaid their analysis of human societies. By providing a thoroughly researched and powerfully argued confirmation of that assumption, Darwin's book completed historical materialism. This was the materialist explanation of the historical character of nature they knew must be possible. As Engels wrote in *Socialism: Utopian and Scientific*:

> Nature works dialectically and not metaphysically . . . she does not move in the eternal oneness of a perpetually recurring circle, but goes through a real historical evolution. In this connection, Darwin must be named before all others. He dealt the metaphysical conception of Nature the heaviest blow by his proof that all organic beings, plants, animals, and man himself, are the products of a process of evolution going on through millions of years.[27]

Nature and Society

A key element of D. A. Stack's claim that Engels tried to "steal the Darwinian mantle" but wasn't really committed to Darwinism, is his assertion that Engels did not make "any meaningful or successful attempt to unite Marxist politics with Darwinian science."[28]

If we accept a very narrow definition of politics, this charge is absolutely true. Engels didn't just fail to propose a political program based on Darwin's science, he explicitly denied that such a program was appropriate.[29]

The idea that the theory of natural selection was an appropriate basis for understanding and governing human societies originated with the English libertarian philosopher Herbert Spencer, the man who originated the phrase "survival of the fittest." He argued that natural selection would eventually produce a perfect society, but only if it had free rein to operate so that the unfit could be eliminated. To that end he opposed public education, compulsory smallpox vaccination, free libraries, workplace safety laws, and even charitable support for the "undeserving poor."

Such views, later labeled "Social Darwinism," were eagerly adopted by defenders of unfettered capitalism. Oil tycoon John D. Rockefeller famously told a Sunday School class in New York City:

> The growth of large business is merely a survival of the fittest. . . . The American Beauty rose can be produced in the splendor and fragrance which bring cheer to its beholder only by sacrificing the early buds which grow up around it. This is not an evil tendency in business. It is merely the working out of a law of nature and a law of God.[30]

Engels was scathing in his rejection of attempts to apply biological laws to human society. In a letter to the Russian socialist Pyotr Lavrov in 1875, he pointed out that the "bourgeois Darwinians," referring to a political current in Germany that claimed to be applying Darwin's views, first claimed that the political concept "survival of the fittest" applied to nature, and then reversed the process:

> All that the Darwinian theory of the struggle for existence boils down to is an extrapolation from society to animate nature of Hobbes's theory of the *bellum omnium contra omnes* [war of all against all] and of the bourgeois-economic theory of competition together with the Malthusian theory of population. Having accomplished this feat . . . these people proceed to re-extrapolate the same theories from organic nature to history, and then claim to have proved their validity as eternal laws of human society. The puerility of this procedure is self-evident, and there is no need to waste words on it.

These political Darwinians, Engels concluded, can be described "firstly as bad economists and secondly as bad naturalists and philosophers."[31]

In 1845, in *The German Ideology*, Marx and Engels had argued that the ability to produce life's needs distinguishes humans from other animals:

Men can be distinguished from animals by consciousness, by religion or anything else you like. They themselves begin to distinguish themselves from animals as soon as they begin to produce their means of subsistence, a step which is conditioned by their physical organization. By producing their means of subsistence they are indirectly producing their material life.[32]

Engels repeated and extended that argument in the late 1870s, in his unfinished book *Dialectics of Nature*:

Let us accept for a moment the phrase "struggle for existence," for argument's sake. The most that the animal can achieve is to collect; man produces, he prepares the means of subsistence, in the widest sense of the words, which without him nature would not have produced. This makes impossible any unqualified transference of the laws of life in animal societies to human society.[33]

Engels was restating a fundamental element of the Marxist view of nature—that different forms and complexities of matter involve different scientific laws. The laws governing the movements of atoms and molecules are not the same as the laws that govern the movements of billiard balls. And, if recent developments in astrophysics are to be believed (the hypothetical existence of dark matter and dark energy, for example), the movements of galaxies may follow still different laws.

The laws that govern inorganic matter also apply to living matter, but they are enhanced and in many respects replaced by biological laws that cannot be reduced to or deduced from Newtonian physics. Similarly, human beings are physical and biological objects, subject to the same physical and biological laws as other animals, but we are also social beings who produce our means of existence, so our lives and history cannot be fully explained by physics and biology.

As Engels wrote, "The conception of history as a series of class struggles is already much richer in content and deeper than merely reducing it to weakly distinguished phases of the struggle for existence."[34]

Darwin's Achievement

The materialist victory in science is one of humanity's greatest achievements. For that reason alone, no matter what his hesitations, delays, or middle-class prejudices, Charles Darwin deserves to be remembered and honored by everyone who looks forward to the ending of superstition and ignorance in all aspects of life.

Darwin was not a political radical; apart from his lifelong opposition to slavery and his involvement in the affairs of the small town where he lived, he seems to have had little interest in political activity or theory. And yet, as the evolutionary biologist Ernst Mayr wrote, "In his scientific works he systematically demolished one after the other of the basic philosophical concepts of his time and replaced them with revolutionary new concepts."[35]

By doing that, Darwin unwittingly contributed to and strengthened the most revolutionary social theories ever developed, the ideas we know today as Marxism.

It is obviously possible, as Paul Heyer points out, to be a Darwinian in biology while rejecting Marxism, but it is not possible to be a consistent Marxist and reject Darwin.

> The reason is basic. Central to Marx's vision is the assumption that nature and history fit together to comprise a totality. Since man emerged from and continues to depend on and transform nature, history as a science will remain incomplete until this foundation is fully comprehended. And no one has contributed more toward this comprehension than Darwin.[36]

The idea that nature has a history, that species come into existence, change, and disappear through natural processes, is just as revolutionary, and just as important to socialist thought, as the idea that capitalism isn't eternal, but came into being at a given time and will one day disappear from the earth.

PART TWO

RESPONDING TO THE ANTHROPOCENE

By the time humans realize that a business-as-usual
approach may not work, the world will be committed to
further decades or even centuries of environmental change.
Collapse of modern, globalized society under uncontrollable
environmental change is one possible outcome.
— WILL STEFFEN, PAUL J. CRUTZEN,
AND JOHN R. McNEILL[1]

CHAPTER 3

A Challenge that Socialists Cannot Ignore

SR: Can you explain the concept of the Anthropocene and its importance for understanding the current climate crisis?

IA: Anthropocene is the proposed name for the present stage of Earth history: a time in which human activity is transforming the entire planet in unprecedented and dangerous ways.

Scientists divide Earth's 4.5-billion-year history into time intervals that correspond to major changes in the conditions and forms of life on Earth. The Pleistocene epoch, which began about 2.5 million years ago, was marked by what are commonly known as ice ages. The glaciers finally retreated about 11,700 years ago, initiating what geologists call the Holocene epoch, characterized by a relatively stable and warm climate. Holocene conditions made agriculture possible, and that in turn laid the basis for the rise and spread of human civilizations.

Adapted from "Entering the Age of Humans," an interview with Ian Angus in *Socialist Review*, May 2016.

As Johan Rockström of the Stockholm Resilience Centre says, Holocene conditions are the only ones that we know for sure are compatible with complex human societies.

In the 1980s a growing number of scientists became concerned that human activity was affecting more than local environments and specific ecosystems; that some changes were affecting the world as a whole. Examples included radioactive fallout from nuclear bomb tests, damage to the ozone layer caused by widely used chemicals, and global climate change caused by greenhouse gas emissions. These concerns led to the launch of the largest and most complex program of international scientific cooperation ever undertaken, the International Geosphere-Biosphere Program (IGBP), which coordinated the efforts of thousands of scientists around the world from 1990 to 2015.

The IGBP's work produced a huge leap forward in scientific understanding of the Earth System, including convincing evidence that human activity is not just disrupting that system, but is doing so in ways that are more extensive and fundamental than anyone had imagined. In 2000, at an IGBP meeting to review its first decade of research, the Nobel Prize–winning chemist Paul Crutzen argued on the basis of that evidence that defining features of the Holocene no longer exist, that Earth has entered a new epoch. He suggested the name Anthropocene, from the Greek word *anthropos*, meaning human being.

Among the most important advances made by the IGBP is a much fuller understanding of the global carbon cycle and its relation to climate change. Scientists have known since the 1800s that carbon dioxide in the atmosphere regulates the global temperature by allowing solar energy to reach Earth as light, while restricting the reflection of heat back into space. If there was a lot more carbon dioxide in the air, Earth would be like Venus—too hot to support life. If there was far less, Earth would be like Mars—too cold for life. But though it was known by the 1960s that atmospheric carbon dioxide was increasing, scientists didn't know if the increase was unusual, and if so, how out of the ordinary it was, and what its effects might be.

Thanks to studies coordinated by the IGBP, we now know that for at least eight hundred thousand years atmospheric carbon dioxide has varied within a strictly limited range, never lower than 180 parts per million in cold times, never higher than 300 ppm in warm times. Over very long periods carbon dioxide has cycled between the atmosphere and the oceans, keeping Earth's temperatures within surprisingly well-defined limits. Today, however, as a direct result of fossil fuel combustion, the concentration is over 400 ppm, and it is growing fast. Human activity has disrupted a complex natural cycle that took millions of years to evolve and stabilize, and that disruption is rapidly changing the state of the planet.

As Crutzen and Will Steffen wrote, "Earth is currently operating in a no-analogue state. In terms of key environmental parameters, the Earth System has recently moved well outside the range of natural variability exhibited over at least the last half-million years. The nature of changes now occurring simultaneously in the Earth System, their magnitudes and rates of change, are unprecedented and unsustainable."[1]

SR: In your book *Facing the Anthropocene: Fossil Capitalism and the Crisis of the Earth System*, you say that climate change is usually seen as a gradual process, but that the reality can be very different. Can you explain what this means?

IA: Science has long known that conditions on Earth have changed over time—that much of the planet has at times been covered in ice, and that areas that are now cold were once tropical—but it was believed that such shifts took place very slowly, over thousands of years or longer. One of the most surprising results of recent research into past climate is the discovery that rapid climate change has been the rule, not the exception. The climate is remarkably sensitive to small changes in the atmosphere and oceans, and rather than gradually warming or cooling, it has tended to lurch from one state to another, in years or decades rather than millennia. As climatologist Wallace Broecker says, Earth's climate is an ornery beast that violently overreacts even to small nudges.

That's particularly relevant today, when greenhouse gas concentrations are not only high but rising more quickly than ever before. That puts unprecedented stress on the climate system, greatly increasing the possibility of runaway climate change, of a relatively quick shift into a completely different climate regime. That could mean, for example, that what we experience today as extreme but rare heat waves could become common or even frequent occurrences. If we cross such a tipping point, ecosystems won't have time to adjust, species won't have time to evolve, and human societies might not have time to adapt. Climate tipping points, by their nature, are impossible to predict; depending on the speed of change, we might not even know we've passed one until well after the point of no return.

That adds urgency to the crisis. We don't know how long we have before climate change goes from dangerous to extremely dangerous, but we know that continuing with business as usual makes such a shift increasingly likely. What we do in this century, and probably in the next few decades, will determine the kind of world our grandchildren will inherit.

SR: You focus on the Second World War as the point at which global environmental change took off. Why was this?

IA: When Crutzen recognized that we are in a transition to a new epoch, he suggested that the Anthropocene may have begun at the time of the Industrial Revolution. That was an important insight. As I show in *Facing the Anthropocene,* fossil capitalism was born when factories and railroads and ships began burning coal on a massive basis. An economic and social system that must grow or die became dependent on fossil carbon to power its growth, and the history of capitalism since then has been marked by ever-increasing use of coal, gas, and oil.

That change began in the Industrial Revolution, but when scientists at the IGBP undertook the detailed work of quantifying the changes in the Earth System that define the Anthropocene, they discovered an unexpected pattern. In almost every case, graphs of long-term trends in the Earth System—atmospheric carbon dioxide, ozone depletion, species extinctions, loss of forests, and so on—show gradual growth from 1750 to about 1950, when a steep increase began. From 1950 to the present the trend lines have gone almost straight up. As the authors of IGBP's synthesis report wrote, "The last 50 years have without doubt seen the most rapid transformation of the human relationship with the natural world in the history of the species."

More and more scientific research has confirmed that conclusion, and most scientists involved in Anthropocene research now argue that even though dependence on fossil fuels began with the Industrial Revolution, the qualitative shift to a new epoch occurred in the second half of the twentieth century. They've dubbed the period since 1950 the "Great Acceleration."

One of the principal things I've tried to do in *Facing the Anthropocene* is to explain the social and economic forces that drove this hockey-stick pattern. Why didn't growth and environmental destruction continue increasing gradually? Or, to put it another way, why did all the trends suddenly accelerate together? When I studied long-term trends in fossil capitalism, including the growth of fossil fuel production and use, the automobilization of Western society, corporate concentration and the rise of monopolies, the mass introduction of synthetic petrochemical-based products, the industrialization of agriculture, and more, I found that the Second World War played a key role in accelerating every single one.

Mainstream economists and historians typically treat wars as anomalies, as interruptions in capitalism's normally peaceful development. In fact, capitalist growth in the twentieth century depended heavily on military production and spending. The most destructive war in human history triggered a radical acceleration of environmental destruction that continues to this day.

SR: How do you respond to those on the left who reject the Anthropocene on the grounds that it blames all of humanity for environmental crisis?

IA: I think it was Rebecca Solnit who wrote that some people on the left seem intent upon finding the cloud around every silver lining. She wasn't talking about responses to the Anthropocene, but her comment certainly applies. Here we have a huge advance in scientific understanding of the Earth System. We have an international discussion about what is to be done, a debate in which almost all participants agree that continuing business as usual is a road to disaster. Ideas that only a few radical environmentalists held until recently are now widely accepted by scientists worldwide, creating the possibility of a powerful science-based challenge to the present social order.

But instead of welcoming that remarkable development, some on the left focus on "proving" that the scientists aren't proper anti-capitalists, that they have illusions about reforming the system, and that some of them see human beings as such as the cause of all environmental destruction, ignoring class and national differences. My response is twofold.

First, I have to wonder whether such critics have actually read what scientists say about this. If they had, they would know that the scientists in the forefront of the Anthropocene project have repeatedly and explicitly rejected any "all humans are to blame" narrative. That's been true from the beginning: in the very first peer-reviewed article on the Anthropocene, Paul Crutzen, the man who invented the word, said clearly that a minority of the world's population was responsible for disrupting the Earth System.[2] That's been a consistent theme of Anthropocene research. The most recent update to the Great Acceleration statistics, published in 2015, says that it has been almost entirely driven by a small fraction of the human population.[3] That's not to say that their social analysis is complete or even adequate, but they are on the right track, and it is shocking that some left critics fail to recognize that.

Second, making such criticisms central to our response builds walls between scientists and the left, when what we should be doing is engaging with scientists, contributing our views and analysis to the global discussion. We need to seize this remarkable opportunity to unite the latest scientific findings with an ecological Marxist analysis in a socio-ecological account of the origins, nature, and direction of the crisis. If the left stays out of the discussion, if we condemn it from the sidelines, we will be leaving Anthropocene science and scientists under the ideological sway of neoliberalism, and we will be irrelevant to the most important scientific development of our time.

SR: In the 1980s, in the face of resistance from chemical companies, the world banned gases that were destroying the ozone layer. What lessons can we take from this for today's crisis?

IA: I devote a chapter of *Facing the Anthropocene* to the ozone crisis because it illustrates so clearly what scientists mean when they say that human activity in the Anthropocene is "overwhelming the great forces of nature." The ozone layer in the upper atmosphere is a vital part of Earth's life support system—it literally makes life possible by blocking deadly ultraviolet radiation. It was only by accident that scientists realized in the 1970s that a family of supposedly harmless synthetic gases known as chlorofluorocarbons, used in refrigerators, aerosol sprays, and other applications, could destroy atmospheric ozone. The giant chemical companies that made CFCs resisted any restrictions or regulations, claiming that there was no actual proof of damage.

It wasn't until the mid-1980s that ground- and satellite-based studies proved beyond doubt that the ozone layer was getting much thinner, much faster than anyone had expected. That led to the Montreal Protocol, an international treaty to phase out CFC production. Even then the largest CFC manufacturer, DuPont, only agreed to phase out production when it was well on the road to developing substitutes that would give it an advantage over its competitors.

One thing we can learn from that experience is that CFC use was reduced not by "putting a price on CFCs" and not by a "CFC cap and trade system" but by a mandatory program of phasing out production, ultimately by an outright ban. That's what worked. The ozone layer won't fully recover for decades, but the process has begun. After two decades of failed attempts to reduce greenhouse gas emissions by "market mechanisms," I think it is past time to apply the Montreal Protocol approach.

An even more important lesson is that corporations will always put their short-term profits first, even if life on Earth is at stake. DuPont and other chemical companies deployed all the arguments and tactics that we've seen from the tobacco and fossil fuel industries to prevent or delay any restrictions on their right to profit from deadly products. If CFCs had been as central to capitalism as a whole as fossil fuels are, the ozone layer might have been gone by now.

SR: In *Facing the Anthropocene*, you argue that fossil fuels are vital to capitalist development, so much so that you call it a "fossil economy." Does this mean that there is no possibility of change while capitalism exists? Do we have to "wait for the revolution" to stop environmental catastrophe?

IA: If you are asking, "Is a permanent and humane solution to the planetary emergency possible within a capitalist framework?" then the answer is certainly no. As I argue in my book, fossil fuels are not an overlay that can be peeled away from capitalism, leaving the system intact. They are embedded in every aspect of the system, in every industry, and there is no prospect of that changing short of an unprecedented global economic collapse.

We know that if fossil fuels continue to be burned the result will be environmentally catastrophic. A small minority of the world's population might survive and even prosper, but most will experience a radical deterioration in their conditions of life. And that's the best outcome. Unless we can replace the present system with one that does not need to "grow or die"—with what I would call ecosocialism—the long-term prospects for humanity are not good.

But this does not mean that nothing can be done short of "the revolution." Capital is not all-powerful; as centuries of experience prove, it can be forced to accept reforms, even substantial ones, if it is faced by a strong enough counterforce. We may not yet be strong enough to stop and reverse capitalism's destructive course, but we can unite the broadest possible range of people to make the political and economic costs of inaction on the environmental crisis unacceptable to the powers that be. In doing so we can win time for Earth and for humanity.

I'll go further: if we don't fight for immediate gains, if we can't build movements that can stop pipelines or fracking, or that can win limits on greenhouse gas emissions, what would make us think that we can succeed in the vastly more difficult task of ending capitalism? It is only by engaging in such campaigns that we can build a broader movement for comprehensive social change.

CHAPTER 4

Ecomodernists and the "Good Anthropocene"

What can lobbyists do when science contradicts their political messages? Some simply deny the science, as many conservatives do with climate change. Others pretend to embrace the science, while ignoring or purging the disagreeable content. That's what the Breakthrough Institute (BTI) is doing with one of the most widely discussed issues in twenty-first-century science, the proposal to define a new geological epoch, the Anthropocene.

BTI has been described as "the leading big money, anti-green, pro-nuclear think tank in the United States, dedicated to propagandizing capitalist technological-investment 'solutions' to climate change."[1] Founded in 2003 by lobbyist Michael Shellenberger and pollster Ted Nordhaus, its philosophy is based on what's known in academic circles as ecological modernization theory—the view that "industrialization, technological development, economic

---

Adapted from Ian Angus, "Hijacking the Anthropocene," *Climate & Capitalism*, May 19, 2015.

growth, and capitalism are not only potentially compatible with ecological sustainability but also may be key drivers of environmental reform."[2]

In BTI's simplified pop version, to which they've assigned the catchier label *ecomodernism*, there is no "may" about it. Their literature consistently couples a professed concern for the environment with rejection of actual pro-environmental policies on the grounds that new technology, growth, and capitalism will solve, and are the only possible solution to, all environmental concerns.

Most notably, BTI opposes efforts to limit greenhouse gas emissions, claiming that investment in nuclear reactors and shale gas will produce all the energy we need, and global warming will wither away as a side effect. "The best way to move forward on climate policy," write Shellenberger and Nordhaus, "is to not focus on climate at all."[3]

As Australian environmentalist Clive Hamilton comments, BTI's founders "do not deny global warming; instead they skate over the top of it, insisting that whatever limits and tipping points the Earth System might throw up, human technology and ingenuity will transcend them."[4]

In 2004, Shellenberger and Nordhaus wrote a notorious pamphlet, *The Death of Environmentalism*. That title wasn't an announcement, it was a goal. They declared their conviction "that modern environmentalism . . . must die so that something new can live."[5] Their organization has worked to achieve that death ever since.

In articles published in *Monthly Review*, Bill Blackwater has exposed the "self-contradictions, simplistic fantasy, and the sheer insubstantiality" of BTI's thought, and John Bellamy Foster has shown that ecological modernization theory involves "a dangerous and irresponsible case of technological hubris [and] a fateful concession to capitalism's almost unlimited destructive powers."[6] In this article, I examine one specific feature of BTI's current activity: its attempt to hijack the Anthropocene, to misrepresent one of the most important scientific developments of our time so that it seems to serve Breakthrough's anti-environ-mental agenda.

Scientists Define the Anthropocene

For scientists, the arrival of a new geological epoch signifies that there has been a qualitative change in the Earth System. For nearly twelve thousand years we have been in the Holocene epoch, but we now face conditions that are as different from that as the Holocene was from the ice-age Pleistocene that preceded it. Paul Crutzen, the Nobel Prize winner who first suggested that such a change had occurred, and Will Steffen, former director of the International Geosphere-Biosphere Program, write:

> The Earth System has recently moved well outside the range of natural variability exhibited over at least the last half million years. The nature of changes now occurring simultaneously in the Earth System, their magnitudes and rates of change, are unprecedented and unsustainable.[7]

The name Anthropocene was proposed to emphasize that the new epoch is driven by a radical change in humanity's relation-ship with the rest of the Earth System, that "global-scale social and economic processes are now becoming significant features in the functioning of the system."[8]

The shift began with the growing use of fossil fuels in the Industrial Revolution, and went into overdrive in the "Great Acceleration" of economic activity, pollution, and environmen-tal destruction in the second half of the twentieth century. Now human activity is "overwhelming the great forces of nature," to the point that if "the institutions and economic system that have driven the Great Acceleration continue to dominate human affairs . . . [then] collapse of modern, globalized society under uncontrol-lable environmental change is one possible outcome."[9]

Foster, Clark, and York describe the Anthropocene as "both a description of a new burden falling on humanity and a recognition of an immense crisis—a potential terminal event in geological evolution that could destroy the world as we know it."[10] Similarly, the editors of *Nature* say it "reflects a grim reality on the ground, and it provides a powerful framework for considering global change and how to manage it."[11]

By contrast, Nordhaus and Shellenberger want us to believe that everything's going to be just fine, that "by 2100, nearly all of us will be prosperous enough to live healthy, free and creative lives." All we need to do is "once and for all embrace human power, technology, and the larger process of modernization."[12]

Foolish environmentalists, they say, may "warn that degrading nonhuman natures will undermine the basis for human civilization but history has shown the opposite: the degradation of nonhuman environments has made us rich." Environmental problems are merely unfortunate side effects of developments that are fundamentally positive for humanity: "The solution to the unintended consequences of modernity is, and always has been, more modernity."[13]

Hijacking a Word, Misrepresenting Science

Given the huge difference in views, it would have been appropriate and honest for BTI to declare how and why it disagrees with the scientists who have identified profound changes in the Earth System and are proposing to declare a new epoch.

Instead, when the word *Anthropocene* started appearing frequently in academic journals and mainstream media, Nordhaus and Shellenberger jumped on the bandwagon and tried to steer it in a direction more congenial to their views. In contrast to scientists they deem to be depressing, pessimistic, and catastrophist, they declared that the Anthropocene isn't a crisis, it's an *opportunity* to build a global technological utopia, in which humanity embraces nuclear power and shale gas, and we all enjoy U.S.-style consumerism forever.

What they offer is a homeopathically diluted Anthropocene, in which the only remaining trace of Earth System science is the fact that the Earth is dominated by human activity. And even that, BTI insists, is neither a recent development nor a matter for concern.

Nordhaus and Shellenberger gave the game away in an article they wrote for *Orion* magazine and reprinted in a BTI-published e-book. After agreeing that humans are "rapidly transforming nonhuman nature at a pace not seen for many hundreds of millions of years," they wrote:

> But the difference between the new ecological crises and the ways in which humans and even prehumans have shaped nonhuman nature for tens of thousands of years is one of scope and scale, not kind.[14]

Read that again. If it's true, then there is no case for declaring a new epoch. There has been no qualitative change, so we are still in the Holocene, still doing what humans have always done, since long before the ice sheets retreated.

Landscape ecologist Erle Ellis, a senior fellow with the Breakthrough Institute, has been arguing for the "scope and scale, not kind" view in the Anthropocene Working Group, the international committee that is evaluating the new epoch from a geological perspective. He supports an *early Anthropocene*, the view that the Anthropocene began not recently but thousands of years ago, when humans first made large-scale changes to landscapes and ecosystems.

Official endorsement of an early date would strengthen the Nordhaus/Shellenberger claim that there is no qualitative break between current and past human impacts on Earth. As Clive Hamilton and Jacques Grinevald write, the early Anthropocene option justifies a business-as-usual understanding of the present:

It "gradualizes" the new epoch so that it is no longer a rupture due principally to the burning of fossil fuels but a creeping phenomenon due to the incremental spread of human influence over the landscape. This misconstrues the suddenness, severity, duration and irreversibility of the Anthropocene leading to a serious underestimation and mischaracterization of the kind of human response necessary to slow its onset and ameliorate its impacts.[15]

BTI's website describes Ellis as "a leading theorist of what scientists increasingly describe as the Anthropocene,"[16] but doesn't mention that his early Anthropocene position, while compatible with BTI's philosophy, has little support among the other scientists involved.

In January 2015, over two-thirds of the Anthropocene Working Group's members endorsed 1945 as the beginning of the Anthropocene, both because the Great Acceleration is a historical turning point for the Earth System, and because there are a number of markers in geological strata, including the remains of nuclear fallout from nuclear weapons tests. The early Anthropocene argument, they write, unduly emphasizes just one aspect of the case for a new epoch:

> The significance of the Anthropocene lies not so much in seeing within it the "first traces of our species" (i.e., an anthropocentric perspective upon geology), but in the scale, significance and longevity of change (that happens to be currently human-driven) to the Earth System.[17]

Ellis, evidently concluding that he had lost the debate, told an editor of the journal *Nature* that he now opposed making any official decision. "We should set a time, perhaps 1,000 years from now, in which we would officially investigate this. . . . Making a decision before that would be premature."[18] That would allow BTI to continue misusing the word, but he seems to have little support: an article in *Science* in 2015, proposing to "avoid the confinement imposed by a single formal designation," had only four signatures, and of them, only Ellis was a member of the Anthropocene Working Group.[19]

Oxymoron Alert

In 2015, Breakthrough invited influential environmental writers to a luxury California resort, all expenses paid, for a two-day seminar on "The Good Anthropocene," an oxymoron that BTI's founders seem particularly fond of. Phrases like "unprecedented and unsustainable" don't appear to meet their approval.

The seminar's message was revealed in *An Ecomodernist Manifesto*, signed by Nordhaus and Shellenberger and sixteen others, all closely associated with BTI. Subtitled *From the Death of Environmentalism to the Birth of Ecomodernism*, it is self-described as "an affirmative and optimistic vision for a future in which we can have universal human development, freedom, and more nature through continued technological and social modernization."[20]

The manifesto extended the oxymoron, promising "a good, or even great, Anthropocene" if only we will reject the "long-standing environmental ideal . . . that human societies must harmonize with nature to avoid economic and ecological collapse."

Yes, you read that right. BTI's pseudo-Anthropocene requires *deliberately expanding the metabolic rift between humanity and the rest of nature into a permanent chasm*. After all, "humans have remade the world for millennia," so more of the same must be good.

A striking feature of all BTI propaganda is the gulf between the concrete problems they admit exist and what Bill Blackwater calls "the daydream quality of their positive solutions."[21] That is clearly on display in their *Ecomodernist Manifesto*, which proposes to solve the pressing problem of climate change with "next-generation solar, advanced nuclear fission and nuclear fusion," technologies that don't exist and won't soon arrive. In the meantime, BTI proposes reliance on hydroelectric dams, which can cause major environmental problems, and on carbon capture and storage, which doesn't exist in any practical form.

Clearly, BTI's "Good Anthropocene" won't arrive before the climate and other essential elements of the Earth System reach tipping points. As Blackwater says, BTI's purported realism is actually "the very height of fantasy," a contemporary form of what radical sociologist C. Wright Mills used to call crackpot realism.

It's Time to Defog

The pundits, politicians, and CEOs whose interests are served by the Breakthrough Institute don't want to be identified with the science deniers of the far right, but neither do they want the radical measures that responding to the real Anthropocene requires. BTI's fantasy of a Good Anthropocene builds the illusion that both objectives are easily achieved. Don't worry, be happy—technological ingenuity will save capitalism from itself.

BTI could have avoided mentioning the Anthropocene, but that would have left a widely discussed concept unchallenged, posing the possibility that public understanding of the state of the Earth System will grow, strengthening the environmentalism that BTI wants to kill. It's far more effective to appropriate the word, to sow confusion by promoting a caricature that has nothing to do with the actual Anthropocene and everything to do with preserving the status quo.

There can be no question about which side the left is on in this conflict. We may not endorse every element of the Anthropocene project, but we must not allow Earth System science to be hijacked and misused by enemies of the environment. It's our responsibility to help blow away Breakthrough's fog of confusion, and work with scientists to stop capitalism's drive to ecological disaster.

✂

UPDATE: In August 2016, the Anthropocene Working Group reported to the International Geological Congress that its members had voted overwhelmingly in favor of a recent date for the beginning of the Anthropocene. There were 28.3 votes for 1950, 1.3 for 1964, 1.3 for three thousand years ago, and none for any other date.

Erle Ellis responded by calling for the AWG to be replaced by a new body that would reopen the question not only of when the Anthropocene began, but even of how a new epoch is to be defined.[22] His appeal continues BTI's campaign to undermine Anthropocene science, while hijacking and redefining the word to fit their political perspective.

CHAPTER 5

The Most Dangerous
Environmentalist Concept?

Environmentalism has generated more than its share of unhelpful ideas for change, including many that would accomplish little or nothing, and some that are downright dangerous. If you had to pick the most dangerous environmental concept of all, you might choose proposals to build tens of thousands of nuclear power plants, or to block all immigration from the Third World, or to impose mandatory sterilization on the world's poorest women, or to spray chemicals into the atmosphere to block solar heat, or to organize guerrilla attacks to destroy energy infrastructure and cause social breakdown, or even to do nothing at all in the hope that some wonderful new technology will be invented. All of these crackpot ideas have been seriously proposed, and any one of them would more likely wreck the earth than save it.

But if none of them seems dangerous enough for you, perhaps you could consider: Anthropocene science.

Previously unpublished; includes material from Ian Angus, "Knocking Down Straw Figures," *International Socialist Review*, Winter 2016–17.

That improbable nomination is made by Jason W. Moore, an associate professor in the Sociology Department at Binghamton University and author of the controversial book *Capitalism in the Web of Life*.[1] In academic papers, blog posts, conference talks, and books, Moore has made it his mission to convince socialists and environmentalists that the Anthropocene is "the most important—and also the most dangerous—environmentalist concept of our times." It is, he writes, "a fundamentally bourgeois concept." What's more, the Anthropocene and the ecological footprint are "tools of the bourgeoisie."[2]

Moore is the inventor of and principal advocate for a concept he calls "world-ecology"—a misleading label since, as he admits, it has nothing to do with ecology as ecologists understand it:

> Some of us have begun to call this way of thinking world-ecological. As is probably clear by now, I don't mean the "ecology of the world." Our ecology is not the ecology of Nature . . . but the ecology of the *oikeios*: that creative, generative, and multilayered relation of life-making, of species and environments.[3]

Judging by the effort he has devoted to attacking it, Moore sees Anthropocene science—or, as he prefers to say, "the Anthropocene argument"—as a major challenge to world-ecology. He says the Anthropocene is "analytically anemic," that it is "a conceptual and historical mess" based on "a neo-Malthusian view of population" and "fanciful historical interpretations."[4] A major achievement of Anthropocene science has been its demonstration that the present global crisis is unprecedented, but in Moore's view, "Where the Anthropocene perspective goes wrong—so very, very wrong—is in its reckoning of the present conjuncture as unique."[5] All of that is in addition to being fundamentally bourgeois, a tool of the bourgeoisie, and the most dangerous environmental concept of our time.

As we'll see, Moore's attempt to put Anthropocene science on trial is long on allegations, but short on evidence.

Knocking Down Straw People

In debating, a straw person argument is one that attacks a position the opponent doesn't actually hold. Rather than deal with the opponent's actual views, the debater substitutes a distorted, exaggerated, or misrepresented position that is easy to defeat (a straw person or scarecrow) and then knocks it down. Whether he intends to or not, that's precisely what Moore does when he writes about the Anthropocene.

Moore's case against Anthropocene science rests on four central arguments, all made of straw:

1. *Wrong agency.* The Anthropocene blames environmental change on "humanity as an undifferentiated whole," rather than on capitalism.[6]
2. *Wrong name.* A name derived from anthropos is inappropriate.
3. *Wrong method.* "The Anthropocene argument . . . is captive to the very thought-structures that created the present crisis. At the core of these thought-structures is Cartesian dualism."[7]
4. *Wrong time.* The Anthropocene treats the Industrial Revolution as "the Big Bang of modernity." This "metabolic fetish" ignores changes that occurred between 1450 and 1750.[8]

Wrong Agency? Moore repeatedly claims that the "Anthropocene argument" attributes environmental change to "humanity as an undifferentiated whole." Anthropocene science, he writes, "reduces the mosaic of human activity in the web of life to an abstract, homogenous humanity," and implies that "capitalism's socio-ecological contradictions are the responsibility of all humans."[9]

The claim that all humans share the blame for environmental problems has been a feature of liberal green thought for generations, so it isn't surprising to find that tired refrain repeated in some articles about the Anthropocene, but it doesn't arise from Anthropocene science, it is added from the outside. Moore provides no evidence, not a single quotation or reference, of such assertions being made in the scientific literature.[10]

Indeed, he ignores many passages that say exactly the opposite.[11] To cite just one example, in 2015, seventeen Earth System scientists signed a paper that states:

> The current levels of the boundary processes, and the transgressions of boundaries that have already occurred, are unevenly caused by different human societies and different social groups. The wealth benefits that these transgressions have brought are also unevenly distributed socially and geographically. It is easy to foresee that uneven distribution of causation and benefits will continue, and these differentials must surely be addressed for a Holocene-like Earth System state to be successfully legitimated and maintained.[12]

Social and economic analysis may not be these scientists' strongest point, but they clearly do not ignore inequalities of wealth and power or say that all humans are responsible for the global crisis.

Wrong Name? This criticism is related to the previous one. Moore objects to the new epoch's name because it is based on the ancient Greek word *anthropos*, which he says means "humanity as an undifferentiated whole."[13] He extends that wrong-name objection to the phrase "anthropogenic global warming," which he describes as "a colossal falsification," because "global warming is not the accomplishment of an abstract humanity, the *Anthropos*." Elsewhere he criticizes the idea of "anthropogenic drivers" for "presuming a fictitious human unity."[14] (He hasn't said whether he also objects to *anthropology*.)

But what matters is not what ancient Greeks meant by *anthropos*, but what the scientific terms anthropogenic and anthropocene mean today. Anthropogenic global warming does *not* mean warming caused by all humans: it distinguishes changes that are caused by human activity from changes that would have occurred even if no humans were involved. Similarly, *Anthropocene* names a planetary epoch that would not have begun in the absence of human activity, *not* one caused by every person on Earth.

Unlike some writers, Moore does not propose an alternative name for the Earth System crisis that scientists have identified. As we'll see, when he uses the word *Capitalocene*, it is not as a name for the new epoch.

Wrong method? In Jason W. Moore's political vocabulary, there is no insult stronger than "dualism," by which he means viewing society as external to nature, rather than as part of it. A substantial part of *Capitalism in the Web of Life* is devoted to his argument that nature/society dualism is capitalism's original philosophical sin, the ideology that made capitalism possible and protects it today:

> The binary Nature/Society is directly implicated in the colossal violence, inequality, and oppression of the modern world.
> . . . The view of Nature as external is a fundamental condition of capital accumulation. . . . This dualism drips with blood and dirt, from its sixteenth-century origins to capitalism in its twilight.[15]

For Moore, dualism is a fatal obstacle to progressive change: "Efforts to transcend capitalism . . . will be stymied so long as the political imagination is captive to capitalism's either/or organization of reality." No "anti-systemic strategy" is possible so long as we remain "firmly encaged within the prison house of the Cartesian binary."[16]

Thus, when he says that "the Anthropocene argument shows Nature/Society dualism at its highest stage of development,"[17] and that it "locates human activity in one box, the rest of nature in another,"[18] he isn't pointing to a minor error. He is saying that Anthropocene science inherently supports capitalism in fundamental ways and that progress will be impossible unless it is cast aside.

This is another straw person. Whether or not one agrees with Moore's concern about nature/society dualism, it is irrelevant in this case because Anthropocene science *explicitly denies* that society and nature are separate entities. That is very clear in this excerpt from the definition of the Earth System in the IGBP's 2004 synthesis report:

> Human beings, their societies, and their activities are an integral component of the Earth System, and are not an outside force perturbing an otherwise natural system. There are many modes of natural variability and instabilities within the System as well as anthropogenically driven changes. By definition, both types of variability are part of the dynamics of the Earth System. They are often impossible to separate completely and they interact in complex and sometimes mutually reinforcing ways.[19]

It's hard to believe that Moore isn't familiar with that definition, since an article he cites several times says the same: "Humans are not an outside force perturbing an otherwise natural system but rather an integral and interacting part of the Earth System itself."[20]

Anthropocene scientists are not saying that the new epoch is caused by human society attacking nature from the outside, but by "the phenomenal growth of the global socio-economic system, *the human part of the Earth System.*"[21] The accusation of dualism is groundless.

Wrong time? A central tenet of Moore's world-ecology concept is that the most important turning point in all history was the birth of capitalism six hundred years ago. Nothing, before or since, can compare to that. The Industrial Revolution, for example, was just "a cyclical phenomenon of capitalism from the sixteenth century."[22]

That conviction—I'm tempted to say obsession—is reflected in the thousands of words Moore has devoted to denouncing what he calls the "dominant Anthropocene argument"—that the new epoch began with the Industrial Revolution in the late eighteenth century.[23] He argues at length that Anthropocene scientists have ignored "the epochal revolution in landscape change that occurred between 1450 and 1750."[24] This proves that they have misdated "the origins of the modern world," and incorrectly answered the question "When and where did humanity's modern relation with nature begin?"[25]

There are two major problems with that argument.

First, the idea that the Anthropocene began with the Industrial Revolution was a *tentative suggestion* made at the beginning of the Anthropocene discussion by Paul Crutzen, but he and most other Anthropocene scientists soon concluded that the qualitative change in society's relationship with and impact on the Earth occurred after the Second World War, in the period they call the Great Acceleration.

Second, it is irrelevant whether "the origins of the modern world" can be found in 1450 or 1780 or some other date, because Anthropocene science does not ask that question or pretend to answer it. Rather than trying to date "the modern world," it studies "fundamental shifts in the state and functioning of the Earth System that are (1) beyond the range of variability of the Holocene, and (2) driven by human activities and not by natural variability."[26] Past environmental changes are important, but Anthropocene science directs its attention to the unprecedented levels of atmospheric CO_2 and artificial nitrogen, species extinction, ocean acidification, sea level rise, holes in the life-protecting ozone layer, and other potentially catastrophic disruptions of the Earth System that are happening *today*. These are major physical disruptions of our planet's life-support systems, not an abstract "modern world."

Moore seems unaware that, as Clive Hamilton points out,

the "Anthropocene" is not a term coined to describe the continued spread of human impacts on the landscape or further modification to ecosystems; it is instead a term describing a *rupture* in the functioning of the Earth System as a whole, so much so that the Earth has now entered a new geological epoch.[27]

Hamilton continues:

> Of course we can understand the Industrial Revolution as the outgrowth of earlier transformations—of mercantile capitalism, colonialism, and transatlantic trade—but to suggest (as Moore does) that the Anthropocene therefore began in the sixteenth century, even though there were no detectable signs of disturbance to the Earth System until the nineteenth century, is to discount scientific facts altogether, which suggests something disquieting about the evolution of the humanities and social sciences.[28]

Moore's criticisms of Anthropocene science are so off-base that it is hard to believe he has read the key scientific papers that define it. The few he cites are mostly examples of what Jeremy Davies calls "the simplest and most sketchily formed version of the concept, the first-draft Anthropocene."[29] It is particularly telling that his bibliographies do not include either of the two most important works on the subject, the International Geosphere-Biosphere Program's 2004 report *Global Change and the Earth System* and the Anthropocene Working Group's 2014 book *A Stratigraphical Basis for the Anthropocene.*

I am reminded of Frederick Engels's complaint that the socialist movement in Germany in his time suffered from a surfeit of people who "write on every subject which they have not studied, and put this forward as the only strictly scientific method . . . persons who give themselves airs about 'science,' of which they 'really never learnt a word.'"[30]

What Is the Capitalocene?

A central feature of Moore's anti-Anthropocene argument is his promotion of the word *Capitalocene.* He has even edited a book titled *Anthropocene or Capitalocene?*

He's not alone in using that word. Marxist historian Andreas Malm and postmodernist philosopher Donna Haraway seem to have coined it independently, although they gave it quite different meanings. Others have since adopted it informally.

Most people who favor the term Capitalocene agree that Earth System scientists have correctly identified a dangerous new stage in planetary history; they just want a name that focuses attention on capitalism's responsibility for the crisis. For example, when Malm says that "a more scientifically accurate designation . . . would be 'the Capitalocene,'" he makes clear that he is referring to the "new geological epoch" that will last far longer than capitalism itself.[31] Another advocate of that position describes the Capitalocene versus Anthropocene debate as "a useful polemical conceit . . . for determining exactly what side you're on regarding capital."[32] I don't think insisting on a name change helps the green-left cause, but I sympathize with the motivation.

As we've seen, Moore says the idea that Earth has entered an unprecedented epoch is "very, very wrong," so for him *Capitalocene* cannot be a name for the new epoch. Indeed, he insists that "the Capitalocene . . . is not an argument about replacing one word with another."[33] Unfortunately for readers seeking clarity, he never tells us his definition of Capitalocene, and what he does say is obscure and often contradictory. In one place, he says the Capitalocene is "the historical era shaped by relations privileging the endless accumulation of capital," which would make it synonymous with capitalism, but in another he insists that "the Capitalocene does not stand for capitalism as an economic and social system."[34]

On four successive pages of one essay, he gives us half a dozen not-quite definitions. The Capitalocene, he writes, is (1) "partly a play on words"; (2) "a *geopoetics*"; (3) "precisely *not* an argument about geological history"; (4) "an argument about thinking ecological crisis"; (5) "world-ecology's rejection of two frames that dominate environmental social science"; and (6) "a key conceptual and methodological move in rethinking capitalism as 'a historically situated complex of metabolisms and assemblages.'"[35]

Evocative and coy, but insubstantial. Surely a concept that is supposed to be the basis for rejecting a major historical development in natural science deserves more concrete explanation!

Moore is equally vague in addressing the concept he rejects. When a writer for *Viewpoint Magazine* asked him to define Anthropocene, Moore replied that there are "two uses of the term":

> One is the Anthropocene as a cultural conversation, the kind of conversation with friends over dinner or at the water cooler. In this sense the Anthropocene has the virtue of posing an important question: how do humans fit within the web of life? But the Anthropocene cannot answer that question, because the very terms of the concept are dualistic. . . .
>
> The Anthropocene argument in its dominant form, on the other hand, is an absurd historical model. It says more or less that everything starts in England in 1800 with steam engines and coal.[36]

A more knowledgeable interviewer would have asked follow-up questions. What about the crisis of the Earth System? The birth of a new and dangerous epoch in Earth history? The violation of critical planetary boundaries? The unprecedented disruption of our planet's life-support systems with potentially catastrophic results? What about climate chaos, mass extinctions, acidified oceans, poisoned rivers, rising seas, and more?

Moore should have been asked: *What about the science?*

But the important questions weren't asked, and Moore got away with reducing the most important scientific development of our time to water-cooler chat and steam engines, omitting everything that actually matters about the Anthropocene.

The Emperor Has No Clothes

As noted above, Moore says that the Capitalocene is "not an argument about geological history." That raises an obvious question about his book, *Anthropocene or Capitalocene?* What can the title possibly mean? The Anthropocene is *a new stage of planetary history* marked by unprecedented and dangerous disruption of the Earth System. If you don't begin with that, you aren't discussing the Anthropocene. If Capitalocene is "not an argument about geological history," then it is irrelevant to the Anthropocene, and his book is comparing apples to bicycles.

The cover of *Anthropocene or Capitalocene?* says that the contributors "challenge the theory and history offered by the most significant environmental concept of our times: the Anthropocene" and "diagnose the problems of Anthropocene thinking."[37] But, as Jeremy Davies, author of *The Birth of the Anthropocene*, makes clear in his review, the emperor has no clothes.

> *Anthropocene or Capitalocene?* has little to say about the first word in its title, and what it does have to say is very poorly informed about the existing literature. . . .
>
> In four of the seven chapters, there's nothing that could reasonably be described as a critique of the Anthropocene argument. The occasional sentence using the word "Anthropocene"? Yes. Though to be honest, there aren't even many of them. But a quotation from, or a recognizable paraphrase of, any book or paper or article that might be said to represent Anthropocene thinking—let alone some such quotation or paraphrase combined with an argument about or against it? No. (And there's not much about the "Capitalocene" either, to be honest.)[38]

In the three chapters that do attempt to address the subject, Moore repeats the straw person arguments discussed above, while

Eileen Crist delivers a neo-Malthusian rant and refers readers to another book in which she advocates global population control. Daniel Hartley raises five supposed objections to Anthropocene science, but three of them refer to peripheral issues in a single scientific paper, and the other two aren't supported by any references at all.[39]

Only Elmar Altvater's essay shows any real acquaintance with Earth System science, with the result that his account of the global crisis contradicts Moore's in major respects. Altvater does like the word *Capitalocene*, but for him it is the "era of fossil energy powering the modern capitalist industrial system"—a view that Moore has elsewhere derided as "fossil fuel fetishism."[40]

As Davies says, *Anthropocene or Capitalocene?* is promoted as "a systematic book-length critique of the Anthropocene from a critical or leftist point of view," but it is nothing of the kind.[41]

Two Anthropocenes?

In his brilliant critique of recent sociological accounts of the natural sciences, the eminent French sociologist Pierre Bourdieu observed that it seems that "any sociologist who feels so inclined will declare himself the bearer of a 'new paradigm,' a 'new' ultimate theory of the world." Such theories are often promoted by asserting "apparently radical positions" that are "calculated to make big waves, especially on American campuses." This, he wrote, leads to "a typical strategy, that of advancing a very radical position . . . before beating a retreat, in the face of criticism, back to banalities."[42]

That "typical strategy" may explain why Moore, rather than replying to or even acknowledging published criticisms of *Anthropocene or Capitalocene?*, has instead tried to change the terms of the debate. After publishing many articles and two books in which he repeatedly denounced the "Anthropocene argument" as such, in October 2016 he told readers of his blog that he has no problem with the "Geological Anthropocene"—that he was only criticizing the "Popular Anthropocene."[43] So far as I can find, he had not used either term in any previous published work.

Moore carried through this terminological change by revising, without notice or explanation to readers, part two of his 2014 article "The Capitalocene," to include his new "Geological versus Popular" binary. For example, this sentence appears in the 2014 version: "The Anthropocene argument is relatively powerless to explain the early modern origins of the relations that enabled the era of humanity as geological agent sometime after 1800." In the October 2016 version the same sentence reads: "The Popular Anthropocene is powerless to explain the early modern origins of the Geological Anthropocene's emergence after 1800."[44]

The words are different, but as we'll see, his anti-Anthropocene argument hasn't changed.

Moore describes the "Geological Anthropocene" as "an argument over stratigraphy . . . an argument over what are going to be the appropriate stratigraphic signals that an archaeologist or an Earth System scientist two million years from now will be able to detect in the global stratigraphic layers."[45] It is, he says condescendingly, "a useful, 'formal concept to the scientific community,'" that has "been eclipsed by the Popular Anthropocene: a way of thinking about the origins and evolution of modern ecological crisis."[46]

Geological/Popular is a strange division. Anyone acquainted with Anthropocene literature knows that it falls into two broad categories, but not those. On one side, there is *Anthropocene science*—hundreds of books, papers, articles, and talks by scientists and science writers whose concern is dangerous global change and the new epoch, and many more technical papers reporting current research. Geology is only one of many sciences involved, because the Anthropocene isn't just a new geological epoch, it is *a crisis of the Earth System*. As a recent paper by leading stratigraphers and Earth System scientists points out, this "major shift in the state of the Earth System was proposed on the basis of direct observations of changes in the Earth System, *without specific reference to evidence in the stratigraphic record*."[47] Moore's

"Geological Anthropocene" reduces Anthropocene science to just one element, and a far larger body of multidisciplinary research and knowledge—most of Anthropocene science, in fact—simply disappears.

The other major component of Anthropocene literature is a mixed bag of essays and books by liberals, critical theorists, deep ecology mystics, ecomodernists, Malthusians, poets, postmodernists, and others, all trying to squeeze the Anthropocene into their particular prejudices and preconceptions. They agree on very little, and their presentations of Anthropocene science range from not bad to truly awful. Some ignore science entirely. Needless to say, it's impossible to find anything resembling a common "popular" perspective in that ball of confusion, let alone one that fits Moore's description:

> The Popular Anthropocene is but the latest of a long series of environmental concepts whose function is to deny the multi-species violence and inequality of capitalism and to assert that the problems created by capital are the responsibility of all humans. The politics of the Anthropocene . . . is resolutely committed to the erasure of capitalism and the capitalogenesis of planetary crisis.[48]

That description repeats a straw person argument Moore has made many times before, with just as little evidence. If his Popular Anthropocene is a real phenomenon that has actually *eclipsed* the scientific account of global change, it should be easy to find and quote many articles, papers, and books about the new epoch that are "resolutely committed" to blaming all of humanity and concealing capitalism's role, but Moore's article doesn't cite even one.

Nor does he explain why, if he has actually been talking about a non-scientific Anthropocene all along, so much of his criticism has been directed at science and scientists.

Indeed, in the very article that claims he isn't criticizing science, Moore launches new attacks on it, writing that there is "a vulgar materialism implicit in many global environmental change studies." He calls the phrase *anthropogenic global warming* "a colossal fabrication." The Anthropocene concept proposed by Paul Crutzen in 2000 is a "quasi-empty signifier." And he denounces Earth System science—the science that has played the leading role in identifying, describing, and explaining the Anthropocene—as "a unity of fragments, an idealist unity that severs the constitutive historical relations that have brought the planet to its present age of extinction."[49]

Like his claims that Anthropocene science is wrong, dangerous, and a tool of the bourgeoisie, such comments attempt to delegitimize Anthropocene science, to warn the left against listening to ideologically suspect scientists. Adding the adjective "popular" hasn't changed that.

The Two Cultures Disease

Chapter 4 discussed the ways in which ideologues from the Breakthrough Institute misrepresent the Anthropocene. Their motives aren't hard to understand: they make a good living by promoting unregulated capitalism and new technology as the solutions to all environmental problems. They are well paid to say what their corporate sponsors want to hear.

It is more difficult to understand why Moore and his co-thinkers, who seem sincerely committed to radical change and even claim some affinity with Marxism, should be so determined to delegitimize the most important scientific developments of our time. Despite their insistence that Anthropocene science is dangerous, they do not show how it might affect or undermine action for environmental and social change, nor do they tell us how their rejection of Anthropocene science might help build more effective red-green movements. They interpret the world in various ways, but show little interest in changing it.

It is tempting to dismiss Moore's criticisms of Anthropocene science as just another example of what philosopher of science

Massimo Pigliucci calls nonsense on stilts—"a hodgepodge of good stuff and sheer nonsense, with no way to separate the chocolate from the manure, so to speak."[50] But Moore is not a crank, and his anti-Anthropocene argument is not an isolated case of science-blindness. It seems, rather, to be a particularly egregious example of a common problem in the academic left. C. P. Snow called it the two cultures, the "gulf of mutual incomprehension" between the humanities and the natural sciences.

The splintering of academia into ever-narrower specialties that compete for prominence and funding has reinforced the two cultures divide. It became much worse in the late twentieth century under the influence of the open hostility toward natural science that characterized postmodernism and social constructivism in the humanities and social sciences. Thankfully, those trends have receded, but the damage they did lingers on.

Exacerbating the gap between the two cultures is the pernicious form of intellectual entrepreneurship known as *academic branding*. It's most often seen in business schools, where careers are made by inventing management fads, but it can be found in all departments.[51] All too often, successful establishment and promotion of an academic brand involves treating other scholars and fields of study not as partners in a common effort to understand the world, but as competing brands that must be denounced and defeated in academic turf wars. As Pierre Bourdieu writes, "Sociology suffers greatly from the fact that the pursuit of distinction at any price . . . encourages an artificial emphasis on differences."[52]

A particularly destructive form of the two cultures disease has been described as "the traditional indifference, disdain, or lordly sense of superiority of humanistic knowledge to the sciences."[53] Most of the essays in *Anthropocene or Capitalocene?*, written by academics in the humanities who think themselves qualified to critique natural science concepts they know little about, display advanced symptoms of that disease. So too does Moore's arrogant dismissal of the "Geological Anthropocene" as a "formal concept" of interest only to scientists, and the absence in his work of any suggestion that radicals have anything to learn from Earth System science and scientists.

Today, many scientists are studying Earth as an integrated socio-ecological system, and a growing number of ecosocialists are working toward a synthesis of ecological Marxism and Earth System science. Unfortunately, some leftist academics are resisting efforts to bridge the two cultures gap: like medieval priests wielding crosses to repel vampires, they are determined to avoid any contact with ideologically suspect science and scientists, lest they be politically contaminated.

This is an academic equivalent of the *political sectarianism* that has long plagued the left. As Karl Marx wrote, a political sect finds its justification for existence "not in what it has in common with the class movement, but in the particular shibboleth distinguishing it from that movement."[54] Differences are inevitable, and debates are an essential part of movement building, but both can be destructive if the differences take priority over collaboration to advance the common cause. Moore's condemnation of Earth System scientists and others "*who refuse to name the system*"—meaning those who don't endorse *his view* of the system—is a disturbing indication that he views protecting his world-ecology brand as more important than protecting the ecology of the world.[55]

A Truly *Dangerous Environmental Concept*

If we must name a "most dangerous environmental concept," then sectarian refusal to engage with Anthropocene science should rank high on the list.

We live in a time when decaying capitalism is destroying our planet's life-support systems. To prevent political changes that would end their destructive rule, powerful corporations and politicians actively promote misinformation about Earth System science. As I write this, the new president of the United States is establishing the most anti-science administration in the country's history. Ignorance and obscurantism are on the march.

But we are also seeing huge advances in scientific understanding of the Earth System and growing awareness that change is urgently needed. Realization that we have entered an unprecedented and dangerous epoch has triggered international discussions in which almost all participants agree that business as usual is a road to planetary disaster. Ideas once held by just a few radical environmentalists are increasingly accepted by scientists worldwide.

Sy Taffel, co-founder of the Political Ecology Research Centre in Aotearoa New Zealand, reminds us that the scientists whose research underpins our understanding of the Anthropocene epoch are facing industry-orchestrated smear campaigns, harassing legal action, and government witch-hunts whose goal is to discredit scientific evidence of the dangers posed by continuing business as usual. Consequently, Taffel says, it is crucial for scholars in the humanities "to stand with our colleagues in the sciences who have consistently brought these issues to public attention, rather than distancing ourselves via an outright rejection of their terminology." If we have differences, "we ought to situate such critique within a collegial discussion with the sciences."[56]

Of course, there are few Marxists or ecosocialists in the ranks of Earth System scientists, but, as historian Dipesh Chakrabarty has wisely noted, although "they are not necessarily anti-capitalist scholars . . . clearly they are not for business-as-usual capitalism either."[57] Indeed, some leading climate scientists are doing more to challenge the system *in practice* than the purportedly left academics who criticize them from safe campus offices.

The possibility of a powerful science-based challenge to the present social order is developing before us, and it would be a tragedy if straw person arguments and fears of ideological contamination were to prevent the left from engaging with scientists and joining the global discussion.

If the left stays out of the discussion, if we condemn it from the sidelines and refuse to learn from science, we will be leaving Anthropocene science and scientists under the ideological sway of neoliberalism, and we will be irrelevant to today's most important scientific developments. If we reject Anthropocene science and deny the new epoch's world-historic importance, we will do lasting damage to both science and radical politics, and undermine our ability to carry through the radical social and geophysical transformations that are so desperately needed in our time.

PART THREE

NUMBERS ARE NOT ENOUGH

Numbers suggest, constrain, and refute; they do not, by themselves, specify the content of scientific theories. Theories are built upon the interpretation of numbers, and interpreters are often trapped by their own rhetoric. They believe in their own objectivity, and fail to discern the prejudice that leads them to one interpretation among many consistent with their numbers.

—STEPHEN JAY GOULD[1]

CHAPTER 6

Tunnel Vision at the Royal Society

The radical ecologist Murray Bookchin once compared populationism to a phoenix, the mythical bird that periodically burns up and is reborn from its own ashes. No matter how often the "too many people" argument is refuted, it always returns, making the same claim that people are breeding too much and consuming too much, devouring the earth like a plague of locusts.[1]

A recent incarnation of the populationist phoenix is *People and the Planet*, a report published in April 2012 by the leading organization of Britain's scientific establishment, the 350-year-old Royal Society. It is more carefully presented and much less hysterical than the tracts produced by population-reduction lobbyists, but the bottom line is the same: people in the South breed too much and people in the North buy too much, and unless we stop all our breeding and buying we will destroy the earth.

───────────────────────────

Adapted from Ian Angus, "The Royal Society's Tunnel Vision on Population and Poverty," *Climate & Capitalism*, May 23, 2012. Postscript adapted from Ian Angus, "Population, Consumption, and Rectangles," *Climate & Capitalism*, December 12, 2012.

This report unintentionally illustrates an aspect of the populationist argument that Simon Butler and I did not discuss in our book, *Too Many People?*—the *tunnel vision* that prevents mainstream demography, and many environmentalists, from seeing beyond symptoms and understanding the causes of the problems they describe.

In her history of twentieth-century demography, anthropologist Susan Greenhalgh calls it "a highly mathematical field, weak on social theory":

> As a field strongly oriented toward generating numbers under the notion that "if you can't measure it, it isn't important," demography resists theoretical constructs that are inherently difficult to operationalize, measure, and analyze as variables. . . .
>
> Unlike other disciplines, where critical traditions live on in Marxist, feminist, postmodernist, and other alternative periodicals, demography allowed no tradition of critical thought to develop. . . . Although some senior members of the field undoubtedly regretted the loss of intellectual vigor that critical perspectives provided, both their own ideology and intellectual predilections, which were liberal, as well as their field's heavy reliance on government and foundation funding, made it difficult for them to create the political space for dissident voices to be heard.[2]

The working group that produced the Royal Society report is overwhelmingly composed of academic demographers, and the result confirms Greenhalgh's words. A narrow focus on numbers, combined with a refusal to consider perspectives and conclusions from analysts whose views don't match the preconceived ideas of mainstream demography, has produced a report that is rich in numbers but sadly lacking in social analysis or understanding.

John Maynard Keynes famously wrote that "even the most practical man of affairs is usually in the thrall of the ideas of some long-dead economist." That's clearly true of the authors of the

Royal Society report. Priding themselves on a hard-nosed commitment to empirical study and statistics, they are nevertheless disciples of long-dead Thomas Malthus, the nineteenth-century economist who blamed the poor for poverty.

The Experts They Didn't Consult

The Royal Society authors' unwillingness to look beyond the narrow confines of their numbers-explain-everything discipline is displayed in their bibliography.

At first glance it's an impressive list of sources. Eighteen pages in all. By my count, 399 separate items. Papers in academic journals. Reports from UN agencies, the World Bank, and various NGOs. Newspaper articles. Previous Royal Society reports. Books published as long ago as 1899 and books that are "in press," which means they haven't been published yet.

But though the bibliography is large, it is limited to a very narrow range of views.

For example, it includes Paul Ehrlich's long and highly critical 1972 review of Barry Commoner's environmental classic *The Closing Circle*, but not the book itself or anything else by Commoner. Similarly, it includes the 1972 Club of Rome report, *The Limits to Growth*, but not Sussex University's 1973 response, *Thinking About the Future*, which argued that the *Limits* computer model was seriously flawed and its data inadequate. In each of these cases, the authors have cited a work that agrees with demography's overpopulation orthodoxy, while omitting a closely related work that questions it.

That bias in favor of populationist sources is consistent. In addition to omitting the works mentioned above, the Royal Society's list of references includes not one book or article by Lourdes Arizpe, Tom Athanasiou, Asoka Bandarage, Murray Bookchin, Paul Burkett, Andrew Feenberg, John Bellamy Foster, Betsy Hartmann, David Harvey, Nicholas Hildyard, Patricia Huang, Patricia Hynes, Joel Kovel, Francis Moore Lappé, Larry Lohmann, Fred Magdoff,

Mahmood Mamdani, Eric Ross, Peter Rosset, Allan Schnaiberg, or Vandana Shiva. And that's a partial list.

The authors listed above disagree on many issues; they include anarchists, feminists, liberals, and various kinds of socialists, but all of them have written extensively about environmental problems, and they all reject the claim that "overpopulation" is a primary cause of those problems, or that reducing birth rates would be an effective solution.

All of them agree that it is impossible to understand the causes and effects of population growth and individual consumption apart from the structure of social and economic relations within which people produce, reproduce, and consume. For example, Frances Moore Lappé and Rachel Schurman argue for a "power structures analysis" that

> stresses the impact on fertility of women's subordination to men, a condition that contributes to the social pressure for many births. But it places this problem within the context of unjust economic structures that deny people realistic alternatives to unlimited reproduction. Within such a framework, rapid population growth is seen to result largely from efforts by the poor to cope, given their powerlessness in the face of the concentrated economic strength of an elite. . . .
>
> From the level of international trade and finance, down to jobs and income available to men as well as women, anti-democratic structures of decision making set limits on people's choices which ultimately influence their reproductive options. . . .
>
> Without such a concept, we believe it is impossible to understand the complex and interacting problems of poverty, hunger, and population, much less act effectively to address them.[3]

Nicholas Hildyard makes a similar argument:

> Discussions of population and food supply that leave out

power relations between different groups of people will always mask the true nature of food scarcity—who gets to eat and who doesn't—and lead to "solutions" that are simplistic, frequently oppressive and that, ultimately, reinforce the very structures creating ecological damage and hunger.[4]

And since the Royal Society authors didn't read Barry Commoner's *The Closing Circle*, they missed his argument against reducing the discussion of environmental problems to population counts and birth control:

> Both the environmental and population crises are the largely unintended result of the exploitation of technological, economic, and political power. Their solutions must also be found in this same difficult arena. . . . There is a temptation to short-circuit the complex web of economic, social, and political issues and to seek direct biological solutions, particularly for the population crisis. I am persuaded that such reductionist attempts would fail.[5]

I am not saying that reading these works would have changed the Royal Society authors' minds, but surely it is not too much to expect that they would at least consider other views.

The Facts They Didn't Include

The Royal Society's tunnel vision goes much beyond not considering other viewpoints. It also leads them to ignore facts that aren't consistent with their only-population-numbers-matter outlook. This can be seen most clearly in the report's case study on the causes of underdevelopment in the Republic of Niger. A realistic account of the causes of poverty in Niger would include at least some of these points:

- The country was occupied and brutally "pacified" by French armed forces between 1900 and 1922. After achieving formal

independence in 1960, Niger was ruled by corrupt governments with close links to France. French advisors and troops
were used repeatedly to suppress opposition.

- As recently as 2004, Anti-Slavery International estimated that
  nearly 9 percent of Niger's population—over 870,000 people—
  were living in slavery.
- 85 percent of government revenue and over 90 percent of the
  country's foreign trade comes from one industry—uranium
  mines controlled by French corporations. Uranium profits go
  overseas, leaving little for the people of Niger.
- The majority of the people are subsistence farmers or nomadic
  herders. Many have been driven off their traditional lands by
  European or Chinese corporations seeking oil and other minerals.
- Most of the country is desert, and all of it is frequently ravaged
  by droughts that have been made worse by climate change.
- In 2005 the International Monetary Fund imposed a "structural
  adjustment program" that included a 19 percent sales tax on
  basic foodstuffs and abolished emergency grain reserves.

This is by no means a comprehensive list of the factors that
make Niger the fourth-poorest country in the world, but it is
indicative of the profound historical, social, economic, and climatic problems facing the nation.

But none of that is mentioned in *People and the Planet*. The
only cause of poverty in Niger the authors consider relevant is too
many babies.

There is no question that Niger's population is growing rapidly.
The country's population is growing 3.5 percent a year, and the Total
Fertility Rate (the number of children, on average, born to each
woman in her lifetime) is 7.6, nearly three times the global average.

Those facts aren't in dispute. But why are birth rates high in
Niger?

Many researchers have found that large families are an economic necessity in very poor countries. Even very young children
can contribute to family income, and when they grow up they

provide essential support to aging parents. As Lappé and Shurman say, rapid population growth reflects "efforts by the poor to cope, given their powerlessness in the face of the concentrated economic strength of an elite."[6]

There is no suggestion in *People and the Planet* that Niger's large families are a rational response to poverty and powerlessness. Rather, citing surveys showing that most women in Niger want large families, the authors conclude that the country suffers from a "strongly pronatalist culture," and they call for a "massive communication effort" to change that.

It's a perfectly circular argument. Problem: Women in Niger have too many babies because they want to have too many babies. Solution: Wise Northern experts should convince women in Niger to want fewer babies, and then they will have fewer babies. Circular and vacuous.

The result of this narrow analysis leads to the conclusion that economic development in Niger requires "sharp reductions in fertility and population growth." The authors don't explain why slowing population growth might be more effective than directly confronting the devastating impacts of a century of imperialist plunder, invasions, civil wars, and climate-induced famines.

They Have a Hammer . . .

The French expression *déformation professionnelle* means "looking at things from the narrow perspective of one's profession." It's similar to Abraham Maslow's law of the instrument: "If the only tool you have is a hammer, it's tempting to treat all problems as nails."

The authors of *People and the Planet* did not avoid that temptation.

Demography is defined as "the study of the statistics of births, deaths, disease, etc. as illustrating the conditions of life."[7] Statistical information is obviously important for anyone who wants to understand the state of people and our planet, and there is a great deal of useful data in the Royal Society's report.

But demographic analysis is just one tool. Without the tools of history and political economy, statistics are just numbers, not explanations. Lacking something like Lappé's "power structures analysis" to put them in context, the Royal Society's population statistics only confuse and mislead.

Their approach leads them to blame the world's poorest women for their own poverty while ignoring the colonial powers, global corporations, and corrupt dictatorships that have grown rich by stealing the country's wealth.

Because they focus narrowly on the breeding habits of the 99 percent, the Royal Society's authors fail to see the qualitatively more destructive effects of the 1 percent's power and greed.

$$\sim$$

POSTSCRIPT: Shortly after I wrote my response to the Royal Society's report, I encountered a variant of the "numbers explain everything" approach to environmentalism. This version uses elementary school geometry.

I first noticed it when an article in *National Geographic Voices* said the problem isn't population growth but the way humans in rich countries get and use energy. The second person to post a comment wrote: "So, per capita consumption. As someone pointed out recently, that's like saying, 'The biggest contributor to the area of a rectangle is not length. It's width.'"[8]

Paul Ehrlich, of *Population Bomb* infamy, has used the same analogy: "Population and consumption are no more separable in producing environmental damage than the length and width of a rectangle can be separated in producing its area—both are equally important."[9]

Robert Engelman of Worldwatch Institute said the same: "Did someone just remark that these impacts don't stem from our population, but from our consumption? . . . It's as though a geometry text were to propound the axiom that it is not length that determines the area of a rectangle, but width."[10]

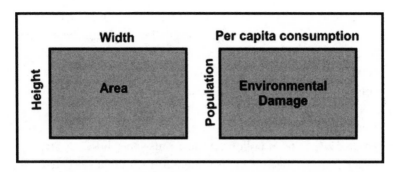

File under *misleading analogies.*

I don't know who said it first, but some populationists evidently think the rectangle analogy is a clever riposte to people who question neo-Malthusian orthodoxy. But—like Malthus's 200-year-old attempt to explain poverty by contrasting the "geometric" increase of population to the "arithmetic" increase of food, this analogy simply doesn't work. It's bad social analysis, and it's bad arithmetic.

To start with, width and height are independent variables— you can change one without changing the other. But you can't determine per capita consumption unless you already know how many people there are: if you change the population, per capita consumption must change. Multiplying them together produces a meaningless result.

More important, the rectangle analogy assumes that there are just two variables, population and individual consumption. In the real world, the production, extraction, and destruction operations of corporations and the military generate far more pollution than all individual consumers put together. So if you must use a geometrical analogy, it will require a complex multidimensional hyper-shape.

Even if we ignore the need for more than two dimensions, and that individual behavior is just a small part of the problem, the rectangle analogy still doesn't work. Height and width are measured in units such as centimeters or inches. Every unit is exactly the same as every other unit, so reducing the height or width by any amount will reduce the area proportionately. Cut the number of centimeters of either dimension in half, and the area will be reduced by 50 percent. It doesn't matter which centimeters you remove—*every one of them* has exactly the same impact on area.

That isn't true of people, because there are huge differences of wealth, class, gender, occupation, geography, and more. A few people cause a lot of pollution; most cause very little. As Stephen Pacala of the Princeton Environmental Institute has pointed out, the three billion poorest people in the world "emit essentially nothing."[11] If every one of the three billion poorest people disappeared today, reducing the "population dimension" by 43 percent, total environmental damage would be essentially unchanged.

On the other hand, a few thousand billionaires, politicians, and generals make daily decisions that cause massive pollution and environmental destruction. Eliminating that tiny minority's power would make a huge difference.

The rectangle analogy is just the latest in a long line of populationist attempts to reduce poverty, hunger, pollution, and climate change to simple numbers. Like its predecessors, this one sounds good but explains nothing.

CHAPTER 7

The Return of the Population Bombers

Promoters of the view that overpopulation is a primary cause of environmental destruction like to pose as underdogs. They regularly claim that there is a taboo on discussing overpopulation, but in the real world their views get far more coverage than the counterargument. Check the environmentalism section of any large bookstore: you'll find dozens of books arguing that overpopulation is destroying the earth. *Too Many People?*, which I wrote with Simon Butler, is one of very few books that challenges overpopulation orthodoxy.[1]

Our book was published in October 2011, which was also, by coincidence, the month chosen by United Nations as the symbolic date when the world's population passed seven billion.

That gave the populationist lobby—the organizations that attribute social and environmental problems to human numbers—an outstanding opportunity to spread their argument through the mass media, and they took full advantage of it. We were treated to a tsunami of articles and opinion pieces blaming the world's

Adapted from a talk by Ian Angus at the Socialism 2012 Conference in Chicago, June 2012.

environmental crises on overpopulation. Numerous environmentalist websites carried articles bemoaning the danger posed by high birth rates.

Global warming, loss of biodiversity, deforestation, food and water shortages, all of these problems and many more were consistently blamed on a single cause: too many people.

In New York's Times Square, representatives of the Committee to Defend Biodiversity handed out condoms in colorful packages depicting endangered animals, and a huge and expensive video billboard warned that human overpopulation is driving species extinct.

In London's busiest Underground stations, electronic billboards paid for by Optimum Population Trust declared that seven billion is ecologically unsustainable.

Otherwise responsible scientific groups joined in promoting the Great Seven Billion Scare. Britain's Royal Society published a major report calling for action to reduce birth rates in poor countries, and an organization that represents 105 global science academies called on the Rio+20 conference to take "decisive action" to reduce population growth.[2]

Populationist ideas are gaining traction in the environmental movement. Sincere activists are once again buying into the idea that overpopulation is destroying the earth, and that what's needed is a radical reduction in birth rates.

Most populationists say they want voluntary birth control programs, but a growing number are calling for compulsory measures. In his best-selling book *The World Without Us*, liberal journalist Alan Weisman says the only way to save Earth is to "limit every human female on Earth capable of bearing children to one."[3]

Another prominent liberal writer, Chris Hedges, writes, "All efforts to staunch the effects of climate change are not going to work if we do not practice vigorous population control."[4]

In *Deep Green Resistance*, Derrick Jensen and his co-writers argue for direct action by small groups, aimed at destroying industry and agriculture and reducing the world's human population by 90 percent or more.

The famous British naturalist Sir David Attenborough tells us that "all environmental problems become harder, and ultimately impossible, to solve with ever more people."[5] Attenborough is a patron of Optimum Population Trust, also known as Population Matters, an influential British group that uses environmental arguments to lobby for stopping immigration.

In the United States, groups such as Californians for Population Stabilization, Carrying Capacity Network, and the grossly misnamed Progressives for Immigration Reform do the same, arguing that immigrants are enemies of the environment.

Anti-immigrant organizations in Canada and Australia have adopted the same strategy. Sadly, we're hearing the same thing from people who actually care about the environment, who aren't just using green arguments as an excuse to bash immigrants.

For example, William Rees, co-creator of the ecological footprint concept, argues that immigration harms the global environment because immigrants adopt the wasteful lifestyles of the wealthy North. He also says that the money immigrants send home to their families will increase consumption in their home countries and so "contribute to net resource depletion and pollution, both local and global."[6] In effect, he says that protecting the environment requires potential immigrants to Stay Away and Stay Poor.

I could cite many more examples. Population growth is once again being identified as the primary cause of environmental destruction, and population reduction is being promoted as the solution. Not just by right-wing bigots, but by sincere though confused environmental activists.

While I was writing this talk, the magazine *Earth Island Journal* was running an online poll on the question "Can you be a good environmentalist and still have children?" The last time I looked, 74 percent had replied, "No." It's a small sample, but indicative.

That's why Simon and I wrote *Too Many People?*—to provide information and arguments that environmentalists and feminists and socialists can use to respond to the many well-meaning but mistaken activists who have adopted populationist ideas and policies.

I'm not going to repeat the arguments in *Too Many People?* here. If you haven't read the book, I hope you will do so soon. Rather I want to offer some historical context, and discuss why the overpopulation argument is so effective and so harmful.

A Conservative Defense of Capitalism

Complaints about overpopulation aren't new, but in the modern era, since the French Revolution, the overpopulation argument has played a specific social and ideological role that we don't find in earlier times.

In good times, the standard capitalist position is that everything is as good as it can be, and everything is getting better. But for over two hundred years, when people have protested the system's massive failures to live up to its promises, the overpopulation argument has been capitalism's fallback position. It provides a biological explanation for social problems, allowing the powers that be to shift blame for human problems away from society onto individual behavior. Sure there are problems, they say, but nothing can be done so long as poor people keep having too many babies.

In the twentieth century, overpopulation theory had its greatest impact between the Second World War and the late 1970s. It began as a response to the revolutionary movements that were then sweeping the Third World, particularly the Chinese Revolution of 1949. The revolutionaries blamed capitalism for poverty and hunger, and promoted socialism as the solution. But if poverty was actually caused by overpopulation, then communism could be defeated by birth control. It's easy to see why that argument appealed to the rich and powerful: it offered a solution that didn't threaten their wealth.

As U.S. President Lyndon Johnson expressed it in 1965, "Less than $5 invested in population control is worth $100 invested in economic growth."[7] That view became the official foreign policy of most wealthy countries. It was implemented directly by Northern government aid agencies, and indirectly through massive fertility control projects organized and funded by the Rockefeller and Ford foundations.

A Conservative Argument Against Social Environmentalism

In the 1960s and 1970s, while overpopulation theory was being used to explain third-world poverty, it was doing double duty as the explanation for the growing environmental crisis. Once again, it provided an explanation and a solution that did not question capitalism, and diverted people away from effective solutions.

I am not suggesting that this was a conspiracy to promote capitalism, or that the theorists of overpopulation didn't believe what they were saying. On the contrary, they were absolutely sincere—they couldn't imagine any alternative to capitalism, so any problems must be the result of individual human failures.

Rachel Carson's wonderful book *Silent Spring*, published in 1962, triggered a new wave of environmental concern and protest. Carson was a left-wing thinker whose book focused attention on the crimes of the chemical industry and the complicity of the governments. The principal causes of ecological degradation, Carson insisted, were "the gods of profit and production." The chief obstacle to sustainability, she wrote, lay in the fact that we live "in an era dominated by industry, in which the right to make a dollar at any cost is seldom challenged."[8]

Carson wasn't alone in offering that kind of social critique. In 1962, in *Our Synthetic Environment*, Murray Bookchin wrote:

> The needs of industrial plants are being placed before man's need for clean air; the disposal of industrial wastes has gained priority over the community's need for clean water. The most pernicious laws of the marketplace are given precedence over the most compelling laws of biology.[9]

And in 1966, Barry Commoner, who had played a leading role in exposing the dangers of nuclear fallout, argued that environmentally destructive technologies had become "deeply embedded in our economic, social, and political structure."[10]

Carson, Bookchin, and Commoner helped initiate a new kind of environmentalism that was rooted in a radical social critique. Their analysis was rejected by the traditional conservationists, the wealthy organizations and individuals whose primary concern was protecting wilderness areas for rich tourists and hunters, not overthrowing capitalism or even protecting human welfare.

In 1968, the oldest and richest of the conservation groups, the Sierra Club, financed the publication, promotion, and wide distribution of a book that was more agreeable to the club's pro-capitalist views. *The Population Bomb,* by Paul Ehrlich, skipped over and ignored all the social complexities and critiques offered by Carson, Bookchin, and Commoner. It contained not one word about corporations, or industrial policies, or markets. Instead it explained all environmental destruction with just three words:

> The causal chain of deterioration is easily followed to its source. Too many cars, too many factories, too much detergent, too much pesticide, multiplying contrails, inadequate sewage treatment plants, too little water, too much carbon dioxide—all can be traced easily to *too many people.*[11]

The Population Bomb was published in 1968, at the very height of the global radicalization, at a time when millions were questioning capitalism, and when a significant part of their questioning focused on the ongoing and increasing destruction of the natural world. The book was heavily promoted by the Sierra Club, by newspapers and television, and by liberal Democrats who correctly saw it as an alternative to the radical views of Carson, Commoner, and Bookchin. It became a huge best-seller, and it played a central role in derailing radical environmentalism.

The impact of Ehrlich's argument could be seen on the official poster that promoted the first Earth Day in 1970. Its main headline was a sentence from Walt Kelly's comic strip *Pogo*: "We have met the enemy and he is us." That often repeated slogan said, in effect, that the threat to the environment wasn't corporations, or capitalism, or profit—it was *people as such*.

At a teach-in during that first Earth Day, Barry Commoner strongly challenged Ehrlich's views. "Pollution," he said, "begins not in the family bedroom, but in the corporate boardroom."[12] He and others carried that argument through the 1970s. Commoner's 1971 book *The Closing Circle* is a classic of socialist environmental thought. I recommend it highly.

But the left lost that battle of ideas. In Murray Bookchin's words:

> Based on my own experience as a very active participant in this momentous period, I can say that if there was any single work that aborted a confluence of radical ideas with public environmental concerns, it was Paul Ehrlich's *Population Bomb*. By the early 1970s, Ehrlich's tract had significantly sidetracked the emerging environmental movement from social critique to a very crude, often odious biologism the impact of which remains with us today.[13]

The transformation happened very quickly. By mid-1971 Zero Population Growth, founded and headed by Ehrlich, had some thirty-six thousand members on four hundred U.S. campuses, making it by far the largest supposedly progressive organization in the country.

Of course, it is impossible to say whether a large anti-capitalist green movement could have been built, but it is clear that the focus on population diverted the energies of tens of thousands of young greens into harmless channels. "Green" became synonymous with population control for the third world and personal behavior change in the North, and most green organizations abandoned the fight for real change.

IN THE 1980S, THE OVERPOPULATION argument seemed to fade away. In part that was because it was obvious to everyone that Ehrlich's predictions of imminent global famine had not materialized. And it was clear that even the largest population control programs had had no impact on Third World poverty. Population control simply did not deliver as promised. What's more, the population control programs financed by U.S. foundations were generating mass resistance: opposition to mass sterilization programs in India played a major role in the 1976 defeat of the government led by Indira Gandhi.

But there was another, deeper cause for the decline of populationism. Put simply, capitalism no longer needed overpopulation theory. The mass radicalization of the 1960s and 1970s was slipping away, and the rise of neoliberalism meant that government intervention in economic and social life, including attempts to control birth rates, was no longer appropriate.

The Reagan government declared that population growth was a non-issue, and slashed spending on family planning programs at home and abroad. George H. W. Bush, who had advocated global population control programs in the 1970s, reversed himself when he ran for president in 1988. Population reduction was out of style in the ruling class.

The Population Bombers Return

Now the population bombers are back, reflecting the renewed need for a non-radical explanation of the environmental crisis. They aren't yet as influential as they were in the 1960s and 1970s, but they are growing, they are being listened to by parts of the ruling class, and they are misleading many new green activists.

If we are to win the battle of ideas this time, we need to understand why the overpopulation argument has been so remarkably successful for so long.

One important factor is its power as a weapon of those who seek to provoke division among the oppressed and hatred of those who are "different." Some of the loudest supporters of populationist policies today are anti-immigrant and racist groups.

But many activists who honestly want to build a better world and are appalled by the racists of the far right are also attracted to populationist arguments. In *Too Many People?*, Simon Butler and I argue that three related factors help to explain why the "too many people" explanation is attractive to some environmentalists.

1. *Populationism identifies an important issue.* Some writers on the left have tried to refute populationism by denying that population growth poses any social, economic, or ecological concerns. Such arguments ignore the fact that human beings require sustenance to live, and that unlike other animals, we don't just *find* the necessities of life, we use the earth's resources to *make* them.

There is a direct relationship between the number of people on earth and the amount of food required to sustain them. That's a fundamental fact of material existence, one that no society can possibly escape. Socialist planning will have to consider population as an important factor in determining what will be produced, and how.

The populationists' error is not that they see the number of people as important, but that they assume that there is no alternative to society's present ways of organizing production and distributing products. In the case of food, they assume that the only way to feed the world's hungry people is to grow more food. Since modern agriculture is ecologically destructive—which it certainly is—feeding more people will cause more destruction, so the only ecologically sound approach is to stop and reverse population growth.

But, as we show in *Too Many People?*, ecologically sound agriculture can produce more than enough food to feed the expected population growth.

Existing food production is in fact more than enough to feed many more people. Just by reducing the food wasted or misused in rich countries to reasonable levels, we could feed billions more people. Or, if population doesn't grow as much as expected, we could help reduce greenhouse gas emissions by reforesting excess farmland.

But such changes will require long-term global planning and coordinated action, which capitalism cannot do.

One of the major tasks facing a post-capitalist society will be to confront and resolve the gross imbalance that capitalism has created between resources and human needs. And that won't be a one-time task. The relationship between human needs and the resources and ecological services needed to meet them will constantly change, and so the need to monitor and adjust will be constant as well. We can't possibly ignore population as a factor in this.

Populationists are right that human numbers must be considered, but they are wrong to blame the imbalance between human needs and resources solely or primarily on human numbers, and they are wrong about the measures needed to solve it.

2. *Populationism reduces complex social issues to simple numbers.* In 1798, Thomas Malthus argued that the imbalance between people and food is a permanent fact of life because population increases geometrically while subsistence only increases arithmetically. He had no evidence for that claim, and history has decisively proven him wrong, but his continuing popularity shows that appeals to numbers can be very effective.

All populationist arguments since then have been rooted in the idea that our numbers determine our fate, that demography is destiny. Hunger, poverty, and environmental destruction are presented as natural laws: surely no reasonable person can argue when the numbers say that population growth is leading us to inevitable disaster.

But even the best population numbers can't explain the environmental crisis, because *quantitative* measures can't take the decisive *qualitative* issues into account. Knowing the number of people in a city or country tells us nothing about the relationships of gender, race, class, oppression, and power that define our connections with one another and our world.

Dr. Lourdes Arizpe, a founding member of the Mexican Academy of Human Rights and former Assistant Director General of UNESCO, poses the issue clearly:

> The concept of population as numbers of human bodies is of very limited use in understanding the future of societies in a global context. It is what these bodies do, what they extract and give back to the environment, what use they make of land, trees, and water, and what impact their commerce and industry have on their social and ecological systems that are crucial.[14]

3. *Populationism promises easy solutions.* The population explanation seems to offer an easy way out of the world's problems, without any need for disruptive social change. Britain's Optimum Population Trust said exactly that in a public statement:

> Providing family planning to everyone who wants it is cheap, effective and popular with users. It is "low hanging fruit" and is much easier than many of the techno-fixes and social changes that others are touting around.[15]

This reminds me of the old joke about a drunk who lost his car keys on First Avenue, but was searching for them on Main Street "because the light is better here." Only major social and economic change can save Earth, so focusing on "easier" birth rates is just as pointless as searching where the light is good instead of where the keys are.

One of the things that makes population reduction seem like an easy solution for many advocates is that it puts the burden of action on *other people.* As the Australian environmentalist Alan Roberts wrote about the previous wave of populationism, over thirty years ago:

It was only too evident that when an ecologist, a population theorist or an economist voiced his alarm at the plague of "too many people," he was not really complaining that there existed too many ecologists, too many population theorists or too many economists: the surplus obviously consisted of less essential categories of the population.[16]

As Simon and I say in our book, for many populationists, "too many people" is code for too many poor people, too many foreigners, and too many people of color.

Confusing Biology and Sociology

Because it separates population growth from its historical, social, and economic context, the population explanation boils down to big is bad and bigger is worse, and its solutions are just as simplistic.

Two centuries ago, the radical essayist William Hazlitt identified the fundamental flaw in Malthus's theory that population growth made poverty inevitable. Malthus, he wrote, viewed the specific social problems and structures of his time as laws of nature.

> Mr. Malthus wishes to confound the necessary limits of the produce of the earth with the arbitrary and artificial distribution of that produce according to the institutions of society or the caprice of individuals, the laws of God and nature with the laws of man.[17]

Modern populationists are more likely to justify the "too many people" argument by reference to the laws of thermodynamics than to the laws of God, but Hazlitt's criticism still applies. Blaming shortages of food and overuse of resources on human numbers confuses sociology with biology; in Hazlitt's words, it treats the "institutions of society" as "laws of God."

The capitalist system is grossly inefficient, inequitable, and wasteful. It cannot create without destroying, cannot survive without mindlessly devouring ever more human and natural resources. Populationist responses to environmental problems search for solutions within a system that is inherently hostile to any solution.

Recognition that the system is itself the problem leads to a different approach, the pursuit of an ecological revolution that will refashion the economy and society, restore and maintain the integrity of ecosystems, and improve human welfare.

Capitalism versus Nature

In *Too Many People?*, Simon Butler and I argue that the fundamental cause of environmental destruction is an economy in which the need to reduce costs and increase profits takes precedence over everything else, including human survival.

Universal access to birth control should be a fundamental human right, but it would not have prevented Shell Oil's massive destruction of ecosystems in the Niger River delta. It would not have halted or even slowed the immeasurable damage that Chevron has caused to rain forests in Ecuador. If the birth rate in Iraq or Afghanistan falls to zero, the U.S. military—the world's worst polluter—will not use a gallon less oil. If every African country adopts a one-child policy, the drive to global climate catastrophe won't slow down one bit.

The "too many people" argument directs the attention and efforts of sincere activists to programs that will not have any substantial effect on the environment, but can be very harmful to human rights. It weakens efforts to build an effective global movement against ecological destruction; it divides our forces, by blaming the principal victims of the crisis for problems they did not cause. Above all, it ignores the massively destructive role of an irrational economic and social system that has gross waste and devastation built into its DNA. The capitalist system and the power of the 1 percent, not population size, are the root causes of today's ecological crisis.

As Barry Commoner said, pollution begins in corporate board-rooms, not family bedrooms. Unless we transform the economy and our society along sustainable lines, we have no hope of securing a habitable planet, regardless of population levels.

Commoner also said that trying to fix the environment by reducing population is like trying to repair a leaky ship by throwing passengers overboard. Instead we should ask if there isn't something radically wrong with the ship.

CHAPTER 8

Did NASA Predict Civilization's Collapse?

I f hundreds of newspaper and online reports are to be believed, scientists at NASA's Goddard Space Agency have proven that Western civilization will collapse unless we radically reduce inequality and shift to renewable resources.

That would be important news if it were true. Is it?

The claim was first made by Nafeez Ahmed, a self-described "investigative journalist and international security scholar" whose blog was hosted by the UK *Guardian*. On March 14, 2014, he described an unpublished "NASA-sponsored study" that, he said, proved that "global industrial civilization could collapse in coming decades due to unsustainable resource exploitation and increasingly unequal wealth distribution."

Ahmed called the paper "a highly credible wake-up call to governments, corporations and business—and consumers—to recognize that 'business as usual' cannot be sustained, and that policy and structural changes are required immediately."[1]

Adapted from Ian Angus, "What Did That 'NASA-Funded Collapse Study' Really Say?," *Climate & Capitalism*, March 31, 2014.

Then the frenzy began. In two days, Ahmed's article was tweeted 6,500 times and shared 100,000 times on Facebook. The mainstream press ran headlines like these:

- "NASA predicts the end of Western civilization" (*New York Post*)
- "The utter collapse of human civilization will be 'difficult to avoid,' NASA funded study says" (*National Post*)
- "NASA-funded report says society is trending toward big collapse" (*Houston Chronicle*)
- "NASA-funded study warns of collapse of civilization in coming decades" (*Times of India*)

Most reporters simply quoted Ahmed's blog, not bothering to read the original study. Liberal and leftish writers did so enthusiastically, while right-wingers fulminated at NASA's anti-capitalist collectivism. If journalists had done their jobs and gone to the source, we might have been spared a lot of scare headlines, unwarranted praise, and outraged denunciations.

Because there is much less here than meets the eye.

Six days after Ahmed's article appeared, NASA publicly denied that it had "solicited, directed or reviewed" the paper, calling it "an independent study by the university researchers utilizing research tools developed for a separate NASA activity."[2]

The paper was signed by three U.S. academics. Lead author Safa Motesharrei was a PhD candidate in mathematics and public policy at the University of Maryland; his co-authors, Eugenia Kalnay and Jorge Rivas, were professors at the University of Maryland and the University of Minnesota, respectively. The only connection to NASA anyone has identified is a general research grant NASA gave to Professor Kalnay's department.

If the original blog post had made that clear, the study wouldn't have attracted much attention; as the headlines show, it was the supposed NASA connection that drew media interest.

Normally I'd ignore a paper like this: many graduate students write papers, some get published, and most are quickly forgotten.

But this one has been widely publicized, so it requires review, if only to understand its argument. After all, if the authors really have proved that Western civilization is on the brink of a collapse that only greater equality and a shift to renewables can prevent, ecosocialists should be eager to publicize it!

Unfortunately, far from being "a highly credible wake-up call," this much-hyped article adds nothing to our understanding of the causes and solutions of the global environmental crisis. The inclusion of egalitarian proposals should not blind environmental activists to its fundamental flaws.

Two somewhat different drafts circulated on the internet: my comments refer to the version dated March 19, 2014, which I'll refer to as "the paper" or by the abbreviated title "Human and Nature Dynamics."

A Universal Model of Social Collapse

The paper's ambitious goal is to propose a mathematical model that explains why societies in general collapse. After listing several dozen complex societies that no longer exist—from the ancient Egyptian, Mesopotamian, and Chinese empires to the Mayan civilization in Central America and Cahokia in the Mississippi Valley—they write that "although many different causes have been offered to explain individual collapses, it is still necessary to develop a more general explanation."

Others have tried to do that. Arnold Toynbee's *A Study of History*, a massive twelve-volume work published between 1934 and 1961, claimed to have identified common factors in the fall of twenty-six major world civilizations. Others have focused on fewer cases: Edward Gibbon filled six large volumes of *The Decline and Fall of the Roman Empire* in an effort to explain just one.

So the authors of "Human and Nature Dynamics" can only be termed ambitious for attempting to identify "a mechanism that is not specific to a particular time period of human history, nor a particular culture, technology, or natural disaster." They are even more ambitious for claiming not just that they have succeeded, but that they can express that mechanism in just four equations: two for population, one for "nature," and one for accumulated wealth.

They accomplish that astonishing feat by accepting, without reservation, the Malthusian claim that the decline of past societies—which they call "collapse," a term they don't define—was caused by population growth and/or mass consumption of limited resources. They identify several sources for that view, most notably Jared Diamond's *Collapse: How Societies Choose to Fail* (Viking, 2005) and William Catton's *Overshoot: The Ecological Basis of Revolutionary Change* (University of Illinois Press, 1980).

They seem unaware that the views of their preferred authors are, to say the least, controversial. Jared Diamond's arguments, for example, were thoroughly debunked by experts on each of the societies he described in the excellent anthology *Questioning Collapse*.[3]

The authors call their four-equation model "HANDY," short for "Human and Nature Dynamics." It is based on the so-called *predator-prey model*, created by mathematicians Alfred Lotka and Vito Volterra in the 1920s to show how interaction between competing species can affect population. The model says, for example, that if there are many rabbits but only a few wolves in an area, the wolf population will rise because food is plentiful, and the rabbit population will fall because they are being eaten. Eventually there will be too many wolves and too few rabbits, so most of the wolves will starve to death. With fewer predators remaining, the rabbit population will increase, and the cycle repeats—lots of rabbits, then lots of wolves, then . . .

That sounds reasonable, but, as population ecologist Daniel Botkin writes, both laboratory experiments and real world observations have shown that "predator and prey do not oscillate as would a Lotka-Volterra set."[4] This suggests Motesharrei and his co-authors are trying to explain fifty centuries of history using formulas that don't even accurately describe a closed system containing just two kinds of animals.

$$\begin{cases} \dot{x}_C = \beta_C x_C - \alpha_C x_C \\ \dot{x}_E = \beta_E x_E - \alpha_E x_E \\ \dot{y} = \gamma y(\lambda - y) - \delta x_C y \\ \dot{w} = \delta x_C y - C_C - C_E \end{cases}$$

Do these four equations explain the collapse of civilizations?

Botkin's book is not listed in the bibliography, and there is no indication that the authors are aware that their fundamental formula may not be valid. They simply adopt it, declaring, "We can think of the human population as the 'predator,' while nature (the natural resources of the surrounding environment) can be taken as the 'prey,' depleted by humans," and proceed.

But nature isn't a thing you can count, so how do you put it into an equation? The authors finesse that problem by replacing it with an imaginary currency they call "eco-dollars," undifferentiated units that somehow reduce nature's variety and complexity to a single number. They don't explain how that could conceivably be done in the real world.

To their credit, the authors recognize that human societies are more complex than the simplistic predator-prey equations allow, so they have modified them to incorporate accumulated wealth (saved eco-dollars) and social inequality. These additions make the model more complex, but they don't fundamentally change it. With humans as "predators" and eco-dollars as "prey," the formulas produce an oscillation: when lots of eco-dollars are available, the human population rises; when the eco-dollars are depleted, the human population falls. Accumulated wealth can delay the population decline and elite overconsumption can accelerate it, but the pattern remains.

Thought Experiments

Having reduced the complexity of human-nature interaction to four equations, the authors devote most of their paper to "thought experiments" about scenarios that represent "three distinct types of societies." In the *egalitarian* scenario, everyone consumes and works equally. In the *equitable* scenario, everyone consumes equally, but some do not work so the rest must work harder. In the *unequal* scenario, non-working elites consume ten to a hundred times as much as commoners, who do all the work.

For each case they calculate a "carrying capacity," the maximum number of people that a given number of eco-dollars can support indefinitely. Although they list Joel Cohen's authoritative work on this subject in their bibliography, they don't mention his conclusion that "the question 'How many people can the earth support?' has no single numerical answer, now or ever."[5]

In each case, depending on how the variables are set, a graph shows whether the carrying capacity will never be exceeded, or will be exceeded intermittently in boom-and-bust cycles, or will be permanently exceeded, leading to total collapse. Although the authors don't say so, it turns out that in the long run the only variables that really matter are total population and total consumption—*because that's how the formulas are defined.*

Their much-quoted statement that for an unequal society "collapse is very difficult to avoid and requires major policy changes, including major reductions in inequality and population growth rates," isn't a political or sociological judgment, it's a mathematical abstraction.

Although the model purports to include all of nature as humanity's "prey," it becomes clear that by "nature" the authors mean "food," because each case of collapse is caused by hunger: either everyone starves because food production declines or commoners starve because the elite eat more than their share. As economist

Christopher Freeman said of an earlier attempt to predict the future with computer models, this is a case of "Malthus in, Malthus out."[6]

Where's the Beef?

One of the most surprising features of the "Human and Nature Dynamics" paper is the absence of empirical data. That's right: *a paper that some say proves the imminent collapse of Western civilization contains no evidence at all.* It fails to show that any past society collapsed from overpopulation and mass or elite consumption, let alone that those are the causes of the environmental crises of the twenty-first century.

In fact, very little such evidence exists.

In 2006, noted anthropologist Joseph A. Tainter, author of *The Collapse of Complex Societies* (Cambridge University Press, 1988), surveyed the subject for the peer-reviewed *Annual Review of Archaeology* (ARA). After summarizing and evaluating all the studies he could find, Tainter wrote:

> When the ARA Editorial Committee invited me to address the topic "Archaeology of Overshoot and Collapse," I assumed I could review only part of a voluminous literature. Although I have extensively read the collapse literature . . . I was surprised to realize that the literature has produced few cases that postulate overshoot of population and/or mass consumption, followed by degradation and collapse. . . . Within the small overshoot literature, many of the most ardent proponents are outside archaeology. . . .
>
> *There does not presently appear to be a confirmed archaeological case of overshoot, resource degradation, and collapse brought on by overpopulation and/or mass consumption.*[7]

Those few sentences fatally undermine the fundamental basis of

all the equations and graphs in "Human and Nature Dynamics."
Tainter's article is not listed in the paper's bibliography.

Back to the Real World

It is indisputable that the existing social order is inflicting massive
harm on humanity and the rest of nature, and the case for radical
social change as the only permanent solution is very strong. The
authors of the HANDY model deserve respect and commenda-
tion for focusing on that, and for including social inequality as
an important factor. My criticism of their work has nothing in
common with the reactionaries and science-deniers who have
loudly condemned them as dangerous radicals.

The problem is not that "Human and Nature Dynamics" is
radical, but that it is not. Radical means going to the root, but this
analysis remains on the surface, dealing with appearances, with
things that can be counted and plugged into formulas. Not society
but population; not nature but eco-dollars; not history and class
struggle but graphs. Ahistorical formulas substitute for investigation of the specific social, economic, cultural, and technological
processes that have brought our particular society to this time of
crisis.

As the British environmentalist and poet Paul Kingsnorth has
written, this is a common failing of mainstream greens: "They
offer up remarkably confident predictions of what will happen if
we do or don't do this or that, all based on mind-numbing numbers cherry-picked from this or that 'study' as if the world were a
giant spreadsheet which only needs to be balanced correctly."[8]

I'm reminded of a letter that Frederick Engels wrote in 1890,
criticizing self-proclaimed Marxists for whom historical materialism "serves as an excuse for not studying history":

> But our conception of history is above all a guide to study, not
> a lever for construction after the manner of the Hegelian. All
> history must be studied afresh; the conditions of existence of

the different formations of society must be examined individually before the attempt is made to deduce them from the political, civil law, aesthetic, philosophic, religious, etc., views corresponding to them.[9]

Engels could have been addressing our modern Malthusians: abstract formulas and assumptions about population and consumption in general cannot substitute for thorough examination of the concrete issues facing specific societies.

Academic economic theory has been described as pure mathematics based on unproven axioms. That's a pretty good description of this paper. Every one of its thought experiments should be preceded by a disclaimer like this:

IF the factors we have identified are the right ones, and
IF there are no countervailing factors, and
IF the formulas are valid and sufficient to explain the process,
THEN these graphs may be worthy of consideration.

If the assumptions aren't true, then it doesn't matter how good the math is or how many graphs are produced, because the entire process is irrelevant to the real world. And *that* is the world we need to understand and change.

PART FOUR

SAVING SPECIES, SAVING OCEANS

The question is whether any civilization can wage
relentless war on life without destroying itself, and without
losing the right to be called civilized.

—RACHEL CARSON[1]

CHAPTER 9

Third World Farming and Biodiversity

n any discussion of biodiversity and species extinction, someone insists that life is a zero-sum game: you can have people and
farming OR wildlife and biodiversity, but not both. Wildlife can
only be preserved by creating preserves from which people are
excluded. The book *Nature's Matrix* presents a convincing antidote
to such views.[1]

Drawing on extensive practical experience with conservation
and agriculture in Central America and the Amazon, combined
with recent research in ecology and agronomy, Ivette Perfecto,
John Vandermeer, and Angus Wright propose a radical new paradigm for conservation, a strategy based on powerful evidence that
preserving biodiversity is inseparable from the growing struggle of peasant farmers for human rights, land, and sustainable
agriculture.

The issue is not how many people there are, but *what the people
do*: some forms of agriculture destroy life, others preserve and
expand it. Only by strengthening the social forces that support

---

Adapted from Ian Angus, "Nature's Matrix: Linking Agriculture, Conservation
and Food Sovereignty," *Climate & Capitalism*, October 17, 2012.

biodiversity-friendly farming can we hope to slow or reverse what's being called the Sixth Extinction; a global species annihilation comparable to the death of the dinosaurs.

A Doomed Strategy

Most conservation groups try to protect biodiversity and limit species extinctions by creating wilderness reserves where human activity is limited or banned. Some groups erect fences and pay armed guards to prevent intrusions, even by people whose ancestors lived on the land for millennia. In their view, what happens outside the reserves is only relevant if it threatens to impinge on the pristine environment where nature is protected from people.

Nature's Matrix argues convincingly that such a focus on creating protected areas is a doomed strategy that actually harms biodiversity, increasing the likelihood of extinctions. This is so for three reasons.

First, most tropical landscapes are neither exclusively untouched nor exclusively farmed. The most common pattern is a complex matrix with fragments of forest separated by a variety of farms. Despite the sincerity and hard work of conservationists, nature preserves will never represent more than a small fraction of biodiversity-rich areas. "Even as the world struggles to protect the few remaining large areas of tropical habitats from further exploitation, we must acknowledge that a large fraction has already been exploited and that perhaps most of the world's biodiversity is located not in those few remaining protected natural areas, but in the far more extensive landscapes in which thousands of islands of natural habitat exist in a matrix of myriad agricultural activities."

Second, within natural habitats, even large ones, local extinctions are normal and inevitable. "Each subpopulation faces a certain likelihood of extinction. Accumulated evidence is overwhelming that extinctions, at this local subpopulation level, are ubiquitous." Local populations must be constantly reinforced by migrations from other areas. "As long as the migration rate balances the extinction rate, the population will survive over the long run." If migration is blocked or seriously inhibited, eventually all of the local populations will disappear and the species will be extinct.

Some kinds of agriculture block such migration, others facilitate it. "There is now little doubt that isolating fragments of natural vegetation in a landscape of low-quality matrix, like a pesticide-drenched banana plantation, is a recipe for disaster from the point of view of preserving biodiversity. It is effectively reducing the migratory potential that is needed if the metapopulations of concern are to be conserved in the long run."

Third, the focus on creating preserves ignores the biodiversity that exists in farmed areas. "Many conservationists think of agriculture as the defining feature of biodiversity loss. The world gets divided into those areas untouched or minimally touched by *Homo sapiens* as contrasted to those areas 'despoiled' by human activity." That view is simply wrong. It is not the existence of agriculture but the *type* of agriculture that determines whether biodiversity is preserved: "Some types of agro-ecosystems contain great amounts of biodiversity while others contain virtually none at all."

What kind of agriculture is most likely to produce a high-quality matrix? The intuitively obvious answer is confirmed by research. "Large-scale production of bananas, sugar cane, tea, technified coffee and cacao, soybeans, cotton, pastures, and others are notorious for their environmental unfriendliness. Small-scale farmers with their usually mixed farming techniques applying, either by conviction or necessity, organic-like and environment-friendly techniques, square far more obviously with the emerging concept of a high-quality matrix."

The authors aren't naive: they recognize that traditional farming isn't always sustainable, and that some of what poses as organic farming is little different from industrial farming. Nor do they suggest that a single model applies everywhere: the most sustainable forms of agriculture depend on the farmers' detailed knowledge of local conditions and their ability to adapt as conditions change.

Nevertheless, experience and research show that a combination of traditional knowledge and the latest agro-ecological research can produce results that are biodiversity-friendly while producing as much food as high-input industrial farms.

A New Paradigm

Unlike most books on environmental issues, *Nature's Matrix* is not limited to a description of the problem and pious wishes that "political will" for change will somehow emerge. Instead the book asks what social forces can be mobilized to change the dominant agricultural model.

Clearly, we cannot expect change to come from the corporations that profit from ever-increasing use of pesticides and artificial fertilizers. Likewise, the giant food growers and processors whose empires depend on monocrop plantations are no friends of biodiversity. Nor are the governments and international agencies that do their bidding.

But a countervailing force is growing. "In many developing countries the small farmers and the landless are organizing and demanding access to land and their right to a decent livelihood. These farmers' organizations, increasingly organized under the banner of food sovereignty, sustainable agriculture and biodiversity conservation, are an integral component of the discourse."

Using examples from Brazil's Landless Workers Movement (MST), Mexico's Zapatistas, and others, the authors of *Nature's Matrix* argue that the fight for biodiversity-friendly farming is inseparably linked to the growing international movement of small farmers against neoliberal agricultural policies and for what the umbrella organization La Via Campesina calls food sovereignty: "the conjoining of the rights of people to consume food to the rights of people to produce their own food."

If the phenomenal rise in rural grassroots social movements continues, it is here that the future of rural landscapes will be determined. Since it is the agriculture in the matrix that is key to the overall conservation agenda, as we have argued, the social movements that form the basis for the future organization of that agriculture become key players in the overall conservation agenda. The Via Campesina, with its philosophy of food sovereignty, agro-ecology and conservation of natural resources, offers an excellent example of where conservation energy should be placed at the present time.

Instead of working with local elites and governments, conservationists must ally with and support the social movements that can actually carry through agrarian revolutions. They must adopt a new paradigm,

> a reorientation of conservation activities, away from a focus on protected areas and towards the sustainability of the larger managed landscape; away from large landowners and towards small farmers; away from the romanticism of the pristine and towards the material quality of the agricultural matrix, nature's matrix. At the centre of this new paradigm is the urgent need for social and environmental justice, without which the conservation of biodiversity ultimately becomes an empty shibboleth.

The conservation strategy outlined in *Nature's Matrix* "gives preference to more progressive political struggles for land rights over traditional lobbying for set-aside reserves of inviolate nature." It rejects blaming species extinctions on birth rates in poor countries, instead identifying the poorest people in those countries as allies. The authors conclude:

> This new paradigm carries with it normative consequences. It suggests that conservation activities need to interact with the rural masses and their social movements more than with wealthy donors. Indeed, we suggest that these new rural

social movements hold the key to real biodiversity conservation. Joining the worldwide struggle of millions of small-scale farmers clamouring for food sovereignty is more likely to yield long-term biodiversity benefits than buying a patch of so-called 'pristine' forest.

Some may call that conclusion utopian, but what's truly unrealistic is the belief that small islands of biodiversity can survive in rising oceans of industrial agriculture. Without an agro-ecological revolution, the Sixth Extinction cannot be stopped, and without the active involvement of mass peasant movements, the agrarian transformation cannot succeed.

Perfecto, Vandermeer, and Wright have written a powerful social and scientific critique of current conservationist policies, and have outlined the basis for a radical alternative. *Nature's Matrix* should be read and heeded by everyone who is concerned about the survival of life on earth.

CHAPTER 10

More Plastic than Fish

I n the 1970s, when most greens were blaming pollution on population growth and personal consumption, socialist environmentalist Barry Commoner showed that neither could account for the radical increase in pollutants since the end of the Second World War. In his best-selling book *The Closing Circle*, he argued that "the chief reason for the environmental crisis . . . is the sweeping transformation of productive technology since World War II."[1] In particular, he pointed to dramatic increases in the production and use of materials not found in nature, synthetics that cannot degrade and so become permanent blights on Earth.

Commoner described natural processes as essentially circular. "In the ecosphere every effect is also a cause: an animal's waste becomes food for soil bacteria; what bacteria excrete nourishes plants; animals eat the plants."[2]

But modern industrial systems are characterized by linear processes: "Machine A always yields product B, and product B, once used, is cast away, having no further meaning for the machine, the product, or the user."

Adapted from Ian Angus, "Plastic Plague," *Monthly Review*, March 2014.

We have broken out of the circle of life, converting its endless cycles into man-made, linear events . . . . Man-made breaks in the ecosphere's cycles spew out toxic chemicals, sewage, heaps of rubbish—testimony to our power to tear the ecological fabric that has, for millions of years, sustained the planet's life.[3]

The petrochemical industry in particular has generated a huge range of materials that nature cannot recycle and reuse. In the first decade after the Second World War, plastics were promoted for and used primarily to make durable products: furniture, tires, automobile and airplane components, and the like. But though plastics are still widely used for long-lasting products, the industry has found its biggest success with throwaways, products specifically designed to be used once and discarded.

Commoner described the trend to disposables in 1971, but he couldn't have known then how bad it would get. When *The Closing Circle* was published, there were no plastic soft-drink bottles, and no one imagined that giant corporations would one day bottle and brand tap water. Today, *72 billion plastic bottles* are produced every year.

Similarly, Commoner wrote before the introduction of plastic grocery bags, which weren't adopted by major supermarket chains until the early 1980s. Today *over 500 billion bags* are made every year. Bottles and bags, together with blister packs, polystyrene tubs, foam peanuts, bubble wrap, Styrofoam trays, candy wrappers, and a multitude of other forms of packaging now account for a third of the plastic produced each year worldwide.[4] It's a bizarre process: products designed to be thrown away are made from materials that never die.

The second of Barry Commoner's famous Four Laws of Ecology, *Everything must go somewhere*, is critically important for materials that degrade extremely slowly, if at all. Some plastic has been burned, some has been recycled, but most, billions and billions of tons of it, remains on Earth and must go somewhere.

In the remarkable book *Plastic Ocean*,[5] Charles Moore reports on the part of the "unstoppable avalanche of nonessentials" that enters the oceans, where it chokes and poisons fish, mammals, and birds, and endangers human life. By turns a memoir, an environmental exposé, and a call to action, *Plastic Ocean* is a dramatic account of what Moore has learned in fifteen years of collecting plastic debris in the Pacific Ocean, studying its effects on marine wildlife, tracing its origins, and campaigning to stop it.

Moore is a rare creature, an activist researcher with the means and determination to work independently of academic and corporate restrictions. Using a family inheritance—ironically, his grandfather was president of Hancock Oil—he founded the Algalita Marine Research Foundation in 1994, hoping "to shorten the distance between research and restoration of the marine environment."

> The "spill, study, and stall" crowd—advocating the hundred-year-old petroleum industry's strategy—demands a science of valueless facts to provide a "complete" mechanistic understanding of a problem before embarking on any solution. Endless, purposeful delay pending perfect "sound science" enforces a form of intellectual sadomasochism driven by the need to preserve profits, not benefit society.

In 1997, while sailing his research boat from Hawaii to California, Moore was initially bemused and then shocked by the amount of plastic litter that floated by, a thousand miles from the nearest land. He later learned that the material he saw was concentrated in the North Pacific Gyre, an area where intersecting currents, prevailing winds, and Earth's rotation combine to produce a slow-moving whirlpool more than twice the size of Texas. There are huge gyres in each of the world's oceans. Any plastic light enough to float that enters the sea, either directly by spills and dumping or carried from land by wind or rivers, is likely to be swept into a gyre, where it will circulate indefinitely, broken by waves and wind into ever-smaller particles:

> Anthropogenic debris—man-made trash, 80–90 percent of it plastic—has broken the reverie of pristine perfection that is the ocean's essence. It's become her most common surface feature.
> . . . Trash has superseded the natural ocean sights, stamping a permanent plastic footprint on the ocean's surface.

Bear in mind that although people have been dumping garbage in oceans for millennia, we've only been throwing plastic away for fifty years. The accumulation of millions of tons of plastic in ocean gyres is powerful confirmation that the nature of garbage changed qualitatively in the last half of the twentieth century.

On the internet, it's easy to find articles that describe the "Great Pacific Garbage Patch" as a floating island of plastic debris, sixty to ninety feet thick. Moore dissociates himself from such fantasies. He describes it as a thin soup that contains untold millions of plastic particles, interspersed with disposable lighters, pieces of fish net, broken buoys, and other objects that haven't broken up yet. If anything, that makes it much more dangerous to the birds, mammals, and fish to which small colored particles suspended in water look like food.

This isn't just about esthetics, the ocean equivalent of roadside trash. Plastic isn't just unsightly, it has made the ocean deadly for its inhabitants. Cases of large animals killed by plastic have been widely publicized—thousands of birds, turtles, seals, and even whales die every year, their throats and guts clogged with indigestible debris—plastic affects the entire animal food chain. As Moore's research has proven, even tiny lanternfish ingest plastic particles, and since they are the main food of tuna, cod, salmon, and shark, many of those particles become part of human diets.

In addition to causing direct physical damage to digestive systems, plastic particles are now known to be an efficient delivery system for toxic chemicals, some from their manufacturing processes, others absorbed as they float by polluted shorelines. Many of these are endocrine disrupters, chemicals that interfere with biological processes, including fetal development. In addition to causing premature death for countless animals, those chemicals ultimately concentrate in the fish that humans eat, contributing to a host of diseases.

This is all well known to science, but there has been little action to stop plastic pollution. The industry has successfully diverted attention away from the production of throwaway plastics to individual consumer behavior; the only actions they support involve cleaning up or recycling products that never should have been made in the first place.

Moore discovered this personally in 2000, when a conference on marine debris rejected his request for discussion of the plastics industry. For the organizers, he realized, "plastics pose a 'handling' problem—that is, a 'people problem,' not a material problem. It's all our fault, in other words."

At a similar conference in 2011, government and industry representatives worked with the organizers to draft what was to be "a plan to lead the way to a debris-free ocean." The official draft didn't even mention plastic until Moore and other activist scientists loudly objected. As Moore told a reporter at the time, trying to clean up the oceans while doing nothing to stop production of disposable plastic is like "trying to bail out a bathtub with the tap still running."

Ultimately, Moore concludes, the problem is a system that puts corporate interests ahead of the environment, even ahead of human survival, "because change is hard, and powerful people and organizations benefit from the status quo. Plastics are a high-stakes game, and those who run it can ill afford to lose control of the playing field. But ridding the oceans of plastic means stopping all plastic inputs—now."

That will only happen if governments "confront industry and put responsibility where it belongs—on plastics producers—not volunteer cleanup crews, taxpayer-supported government agencies, and NGOs."

Unfortunately, despite his indictment of the plastics industry, his rejection of solutions that blame consumers, and his critique of green organizations that ignore "the core issue: the ever-growing volume of plastic products and packaging across the world," Moore stops short of suggesting a strategy for stopping the plastic plague. No one could expect him to provide a detailed roadmap, but his concluding chapter doesn't offer much more than a vague hope that people will demand change.

He argues that "the science and technology that will liberate us from pollution and meaningless labor are available" and predicts that a future generation will "unplug itself from the economy of constant input and output of junk," but he doesn't address the critical question: How do we get there from here? Which battles can be won in the short term and which ones will require radical social change? What kinds of economic and social changes are needed?

Above all, how can we build movements that are strong enough to overcome the power of the giant corporations whose profits depend on pollution?

The book's failure to address such questions isn't really surprising. Many books on environmental problems are published, but few go beyond exposé to explain why, in Commoner's words, "the present system of production is self-destructive; the present course of human civilization is suicidal."[6] Even fewer take the giant step of proposing practical steps toward the ecological revolution that is clearly needed.

It's disappointing that *Plastic Ocean* doesn't take that step because most of the book seems to be leading toward a call for system change. Perhaps this is one more proof that in these times it is easier to imagine the end of the world than the end of capitalism.

Whatever the reason, and despite that weakness, *Plastic Ocean* deserves a wide audience. It is a fine example of what used to be called "people's science." It's as dramatic as many novels, it's accessible to readers with no background in science, and it's an important contribution to broad awareness of the ever-growing conflict between our systems of production and the circle of life.

∽

UPDATE: The original title of this article was "Plastic Plague." I changed it for this book, in recognition of the World Economic Forum's finding that by 2050, if business as usual continues, the plastics in the world's oceans will outweigh all the fish combined.[7]

PART FIVE

TOWARD AN ECOLOGICAL CIVILIZATION

There can be no way out of these
ultimately suicidal contradictions—which are
inseparable from the imperative of endless capital-expansion,
irrespective of the consequences—without radically changing
our mode of social metabolic reproduction. This demands
adopting the responsible and rational practices of the only
viable economy—an economy oriented by human
need, instead of alienating, dehumanizing,
and degrading profit.
—ISTVÁN MÉSZÁROS[1]

CHAPTER 11

The Myth of "Environmental Catastrophism"

Between October 2010 and April 2012, over 250,000 people, including 133,000 children under five, died of hunger caused by drought in Somalia. Millions more survived only because they received food aid. Scientists at the UK Met Office have shown that human-induced climate change made this catastrophe much worse than it would otherwise have been.[1]

This is only the beginning: the UN's 2013 *Human Development Report* says that without coordinated global action to avert environmental disasters, especially global warming, the number of people living in extreme poverty could increase by up to three billion by 2050.[2] Untold numbers of children will die, killed by climate change.

If a runaway train is bearing down on children, simple human solidarity dictates that anyone who sees it should shout a warning, and that anyone who can should try to stop the train. It is difficult

Adapted from Ian Angus, "The Myth of 'Environmental Catastrophism,'" *Monthly Review*, September 2013.

to imagine how anyone could disagree with that elementary moral imperative.

And yet some do. Increasingly, activists who warn that the world faces unprecedented environmental danger are accused of catastrophism—of raising alarms that do more harm than good. That accusation, a standard feature of right-wing attacks on the environmental movement, has recently been advanced by some left-wing critics as well. While they are undoubtedly sincere, their critique of so-called environmental catastrophism does not stand up to scrutiny.

From the Right . . .

The word *catastrophism* originated in nineteenth-century geology, in the debate between those who believed all geological change had been gradual and those who believed there had been episodes of rapid change. Today, the word is most often used by right-wing climate-change deniers for whom it is a synonym for alarmism.

- The Heartland Institute: "Climate Catastrophism Picking Up Again in the U.S. and Across the World"[3]
- A right-wing German blog: "The Climate Catastrophism Cult"[4]
- The Australian journal *Quadrant*: "The Chilling Costs of Climate Catastrophism"[5]

Examples could be multiplied. As environmental historian Franz Mauelshagen writes, "In climate denialist circles, the word 'climate catastrophe' has become synonymous with 'climate lie,' taking the anthropogenic greenhouse effect for a scam."[6]

Those who hold such views like to call themselves climate change skeptics, but a more accurate term is climate science deniers. Though there are uncertainties about the speed of change and its exact effects, there is no question that global warming is driven by greenhouse-gas emissions caused by human activity, and that if business as usual continues, temperatures will reach levels higher than any seen since before human beings evolved. Those who disagree are not skeptical, they are denying the best scientific evidence and analysis available.

The right labels the scientific consensus as catastrophism to belittle environmentalism, and to stifle consideration of measures to delay or prevent the crisis. The real problem, they imply, is not the onrushing train, but the people who are yelling "Get off the track!" Clearing the track would disrupt business as usual, and that is to be avoided at all costs.

. . . and from the Left

Until recently, catastrophism as a political expression was pretty much the exclusive property of conservatives. When it did occur in left-wing writing, it referred to economic debates, not ecology. But in 2007 two quite different left-wing voices almost simultaneously adopted catastrophism as a pejorative term for radical ideas about climate change they disagreed with.

The most prominent was the late Alexander Cockburn, who in 2007 was writing regularly for *The Nation* and co-editing the website *CounterPunch*. To the shock of many of his admirers, he declared that "there is still zero empirical evidence that anthropogenic production of CO_2 is making any measurable contribution to the world's present warming trend," and that "the human carbon footprint is of zero consequence."[7] Concern about climate change was, he wrote, the result of a conspiracy "between the Greenhouse fearmongers and the nuclear industry, now largely owned by oil companies."[8]

Like critics on the right, Cockburn charged that the left was using climate change to sneak through reforms it could not otherwise win: "The left has bought into environmental catastrophism because it thinks that if it can persuade the world that there is indeed a catastrophe, then somehow the emergency response will lead to positive developments in terms of social and environmental justice."[9]

Cockburn's assault on environmental catastrophism was shocking, but his arguments added nothing to the climate debate. They were the same criticisms we had long heard from right-wing deniers, albeit with leftish vocabulary.

That was not the case with Leo Panitch and Colin Leys. These distinguished Marxist scholars are by no means deniers. They began their preface to the 2007 *Socialist Register* by noting that "environmental problems might be so severe as to potentially threaten the continuation of anything that might be considered tolerable human life" and insisting that "the speed of development of globalized capitalism, epitomized by the dramatic acceleration of climate change, makes it imperative for socialists to deal seriously with these issues now."[10]

But despite that clear statement of the severe dangers posed by climate change, they continued: "Nonetheless, it is important to try to avoid an anxiety-driven ecological catastrophism, parallel to the kind of crisis-driven economic catastrophism that announces the inevitable demise of capitalism." Capitalism's "dynamism and innovativeness," they argued, might enable it to use "green commerce" to escape environmental traps.

The problem with the Panitch-Leys argument is that the threat of ecological catastrophe is not "parallel" to the view that capitalism will destroy itself. The desire to avoid the kind of mechanical determinism that has often characterized Marxist politics, where every crisis was proclaimed to be the final battle, led these thoughtful writers to confuse two very different kinds of catastrophe.

The idea that capitalism will inevitably face an insurmountable economic crisis and collapse is based on a misunderstanding of Marxist economic theory. Although economic crises are endemic to capitalism, the system can always continue. Only class struggle, only a social revolution, can overthrow capitalism and end the crisis cycle.

Large-scale environmental damage is caused by our destructive economic system, but its effect is the potentially irreversible disruption of essential natural systems. The most dramatic example is global warming: Earth is now the warmest it has been for thousands of years, and temperatures are rising much faster than at any time since the last Ice Age. Arctic ice and the Greenland ice sheet are disappearing faster than predicted, raising the specter of flooding in coastal areas where more than a billion people live. Extreme weather events, such as giant storms, heat waves, and droughts, are becoming ever more frequent. So many species are going extinct that many scientists call it a mass extinction event, comparable to the time 66 million years ago when 75 percent of all species, including the dinosaurs, were wiped out.

As the editors of *Monthly Review* wrote in reply to *Socialist Register*, if these trends continue, "we will be faced with a different world—one in which life on the planet will be massively degraded on a scale not seen for tens of millions of years."[11] To call this "anxiety-driven ecological catastrophism, parallel to . . . economic catastrophism" is to equate an abstract error in economic theory with some of the strongest conclusions of modern science.

A New "Catastrophism" Critique

An essay provocatively titled "The Politics of Failure Have Failed" offers a different and more sweeping left-wing critique of "environmental catastrophism." Author Eddie Yuen is associated with the Pacifica radio program *Against the Grain*, and is on the editorial board of the journal *Capitalism Nature Socialism*.

His paper is part of a broader effort to define and critique a body of political thought called *Catastrophism*, in a book by that title.[12] In the book's introduction, Sasha Lilley offers this definition:

Catastrophism presumes that society is headed for a collapse, whether economic, ecological, social, or spiritual. This collapse is frequently, but not always, regarded as a great cleansing, out of which a new society will be born. Catastrophists tend to believe that an ever-intensified rhetoric of disaster will awaken the masses from their long slumber—if the mechanical failure of the system does not make such struggles superfluous. On the left, catastrophism veers between the expectation that the worse things become, the better they will be for radical fortunes, and the prediction that capitalism will collapse under its own weight. For parts of the right, worsening conditions are welcomed, with the hope they will trigger divine intervention or allow the settling of scores for any modicum of social advance over the last century.

A political category that includes both the right and the left—and that encompasses people whose concerns might be economic, ecological, social, or spiritual—is, to say the least, unreasonably broad. It is difficult to see any analytical value in a definition that lumps together anarchists, fascists, Christian fundamentalists, right-wing conspiracy nuts, pre-1914 socialists, peak-oil theorists, obscure Trotskyist groups, and even Mao Zedong.

The definition of catastrophism became even more problematic in Yuen's essay.

One of These Things Is Not Like the Others

Years ago, the children's television program *Sesame Street* would display four items—three circles and a square, three horses and a chair, and so on—while someone sang, "One of these things is not like the others, One of these things just doesn't belong. . . ." I thought of that when I read Yuen's essay.

While the book's scope is broad, most of it focuses, as Yuen writes, on "instrumental, spurious, and sometimes maniacal versions of catastrophism—including right-wing racial paranoia, religious millenarianism, liberal panics over fascism, leftist fetishization of capitalist collapse, capitalist invocation of the 'shock doctrine' and pop culture cliché."

But as Yuen admits in his first paragraph, environmentalism is a very different matter, because we are in "what is unquestionably a genuine catastrophic moment in human and planetary history. . . . Of all of the forms of catastrophic discourse on offer, the collapse of ecological systems is unique in that it is definitively verified by a consensus within the scientific community. . . . It is absolutely urgent to address this by effectively and rapidly changing the direction of human society."

If the science is clear, if widespread ecological collapse unquestionably faces us unless action is taken, why is this topic included in a book devoted to criticizing false ideas? Does it make sense to use the same term for people who fear an imaginary train crash and people who are trying to prevent a real crash?

The answer, although he does not say so, is that Yuen is using a different definition than the one Lilley gave in her introduction. Her version used the word for the belief that some form of catastrophe will have positive results—that capitalism will collapse from internal contradictions, that God will punish all sinners, that peak oil or industrial collapse will save the planet. Yuen uses the same word for the idea that environmentalists should alert people to the threat of catastrophic environmental change and try to mobilize them to prevent or minimize it.

Thus, when Yuen refers to "a shrill note of catastrophism" in the work of James Hansen, perhaps the world's leading climate scientist, he is not challenging the accuracy of Hansen's analysis, but only the "narrative strategy" of clearly stating the probable results of continuing business as usual.

Yuen insists that "the veracity of apocalyptic claims about ecological collapse are separate from their effects on social, political, and economic life." Although "the best evidence points to cascading environmental disaster," it is in his view self-defeating to tell people that. He makes two arguments, which we can label practical and principled.

His *practical* argument is that by talking about "apocalyptic scenarios" environmentalists have made people more apathetic, less likely to fight for progressive change. His *principled* argument is that exposing and campaigning to stop tendencies toward environmental collapse has "damaging and rightward-leaning effects"—it undermines the left, promotes reactionary policies, and strengthens the ruling class.

He is wrong on both counts.

The Truth Shall Make You Apathetic?

In Yuen's view, the most important question facing people who are concerned about environmental destruction is: "What narrative strategies are most likely to generate effective and radical social movements?"

He is vague about what "narrative strategies" might work, but he is very firm about what does not. He argues that environmentalists have focused on explaining the environmental crisis and warning of its consequences in the belief that this will lead people to rise up and demand change, but that this is a fallacy. In reality, "once convinced of apocalyptic scenarios, many Americans become more apathetic."

Given such a sweeping assertion, it is surprising to find that the only evidence Yuen offers is a news release describing one academic paper, based on a U.S. telephone survey conducted in 2008, that purported to prove that "more informed respondents both feel less personally responsible for global warming, and also show less concern for global warming."[13]

Note first that being "more informed" is not the same as being "convinced of apocalyptic scenarios" or being bombarded with "increasingly urgent appeals about fixed ecological tipping points." On the face of it, this study does not appear to contribute to our understanding of the effects of "catastrophism."

What's more, reading the original paper reveals that the people described as "more informed" were self-reporting. If they said they were informed, that was accepted, and no one asked if they were listening to climate scientists or to conservative talk radio. That makes the paper's conclusion meaningless.

Later in his essay, Yuen correctly criticizes some environmentalists and scientists who "speak of 'everyone' as a unified subject."

But here he accepts as credible a study that purportedly shows how all Americans respond to information about climate change, regardless of class, gender, race, or political leanings.

The problem with such undifferentiated claims is shown in a 2011 study that examined the impact of Americans' political opinions on their feelings about climate change. The study found that liberals and Democrats who reported being well informed were more worried about climate change, while conservatives and Republicans who reported being well informed were less worried.[14] Obviously the two groups mean very different things by "well informed."

Even if we ignore that, the study Yuen cites is a one-time snapshot. It does not tell us what radicals really need to know, which is how things are changing. For that, a more useful survey is one that scientists at Yale and George Mason University have conducted seven times since 2008 to show shifts in U.S. public opinion.[15] Based on answers to questions about their opinions, respondents are categorized according to their attitude toward global warming. The surveys reveal that:

- The number of people identified as "disengaged" or "cautious"—those we might call apathetic or uncertain—has varied very little, accounting for between 31 and 35 percent of the respondents every time.
- The categories "dismissive" or "doubtful"—those who lean toward denial—increased between 2008 and 2010. Those groups then shrank back almost to the 2008 level.
- In parallel, the combined "concerned" and "alarmed" groups shrank between 2008 and 2010, but then largely recovered. In September 2012—before Hurricane Sandy!—there were more than twice as many Americans in these two categories as there were in dismissive/doubtful.

Another study, published in the journal *Climatic Change*, combined seventy-four independent surveys conducted between 2002 and 2011 to create a Climate Change Threat Index (CCTI), a measure of public concern about climate change, and evaluated how it has changed in response to events. The study found that public concern about climate change reached an all-time high in 2006–7, when the Al Gore documentary *An Inconvenient Truth* was seen in theaters by millions of people and won an Academy Award. The authors concluded: "Our results . . . show that advocacy efforts produce substantial changes in public perceptions related to climate change. Specifically, the film *An Inconvenient Truth* and the publicity surrounding its release produced a significant positive jump in the CCTI."[16]

This directly contradicts Yuen's view that more information about climate change causes Americans to become more apathetic. There is no evidence of a long-term increase in apathy or decrease in concern, and when scientific information about climate change reached millions of people, the result was not apathy but a substantial increase in support for action to reduce greenhouse gas emissions.

"The Two Greatest Myths"

Yuen says environmentalists have deluged Americans with catastrophic warnings, and this strategy has produced apathy, not action. Writing of establishment politicians who make exactly the same claim, noted climate-change analyst Joe Romm says, "The two greatest myths about global warming communications are 1) constant repetition of doomsday messages has been a major, ongoing strategy and 2) that strategy doesn't work and indeed is actually counterproductive!" Contrary to liberal mythology, the North American public has not been exposed to anything even resembling the first claim. Romm writes:

> The broad American public is exposed to virtually no doomsday messages, let alone constant ones, on climate change in popular culture (TV and the movies and even online). . . . The major

energy companies bombard the airwaves with millions and millions of dollars of repetitious pro-fossil-fuel ads. The environmentalists spend far, far less money. . . . Environmentalists when they do appear in popular culture, especially TV, are routinely mocked. . . . It is total BS that somehow the American public has been scared and overwhelmed by repeated doomsday messaging into some sort of climate fatigue.[17]

Similarly, a British study describes coverage of climate change in newspapers there as "lamentably thin," a problem exacerbated by the fact that much of the coverage consists of "worryingly persistent climate denial stories." The author concludes drily: "The limited coverage is unlikely to have convinced readers that climate change is a serious problem warranting immediate, decisive and potentially costly action."[18]

Given Yuen's concern that Americans do not recognize the seriousness of environmental crises, it is surprising how little he says about the massive fossil-fuel-funded disinformation campaigns that have confused and distorted media reporting. I can find just four sentences on the subject in his 9,000-word text, and not one that suggests denialist campaigns might have helped undermine efforts to build a climate change movement.

On the contrary, Yuen downplays the influence of "the well-funded climate denial lobby," by claiming that "far more corporate and elite energy has gone toward generating anxiety about global warming," and that "mainstream climate science is much better funded." He provides no evidence for either statement.

Of course, the fossil-fuel lobby is not the only force working to undermine public concern about climate change. It is also important to recognize the impact of liberals' consistent unwillingness to confront the dominant forces in U.S. capitalism, and of the craven failure of mainstream environmentalist groups and NGOs to expose and challenge the Democratic Party's anti-environmental policies.

With fossil-fuel denialists on one side, and pale-green NGOs on the other, activists who want to get out the truth have barely been heard. In that context, it makes little sense to blame environmentalists for sabotaging environmentalism.

The Truth Will Help the Right?

Halfway through his essay, Yuen abruptly changes direction, leaving the practical argument behind and raising his principled concern. He now argues that what he calls catastrophism leads people to support reactionary policies and promotes "the most authoritarian solutions at the state level." Focusing attention on what he agrees is a "cascading environmental disaster" is dangerous because it "disables the left but benefits the right and capital." He says, "Increased awareness of environmental crisis will not likely translate into a more ecological lifestyle, let alone an activist orientation against the root causes of environmental degradation. In fact, right-wing and nationalist environmental politics have much more to gain from an embrace of catastrophism."

Yuen says that many environmentalists, including scientists, "reflexively overlook class divisions," and thus do not realize that "some business and political elites feel that they can avoid the worst consequences of the environmental crisis, and may even be able to benefit from it." Yuen apparently thinks those elites are right. Although the insurance industry is understandably worried about big claims, he says, "the opportunities for other sectors of capitalism are colossal in scope."

He devotes much of the rest of his essay to describing the efforts of pro-capitalist forces, conservative and liberal, to use concern about potential environmental disasters to promote their own interests, ranging from emissions trading schemes to military expansion to Malthusian attacks on the world's poorest people. "The solution offered by global elites to the catastrophe is a further program of austerity, belt-tightening, and sacrifice, the brunt of which will be borne by the world's poor."

Some of this is overstated. His claim that "Malthusianism is at the core of most environmental discourse," reflects either a very

limited view of environmentalism or an excessively broad definition of Malthusianism. And he seems to endorse David Noble's
bizarre theory that public concern about global warming has been
engineered by a corporate conspiracy to promote carbon-trading
schemes.[19] Nevertheless he is correct that the ruling class will do
its best to profit from concern about climate change, while simultaneously offloading the costs onto the world's poorest people.

But who is he arguing with? This book says it aims to "spur
debate among radicals," but none of this is new or controversial
for radicals. The insight that the interests of the ruling class are
usually opposed to the interests of the rest of us has been central to
left-wing thought since before Marx was born. Capitalists always
try to turn crises to their advantage no matter who gets hurt, and
they always try to offload the costs of their crises onto the poor
and oppressed.

What needs to be shown is not that pro-capitalist forces are
trying to steer the environmental movement into profitable channels, and not that many sincere environmentalists have backward
ideas about the social and economic causes of ecological crises.
Radicals who are active in green movements know those things
perfectly well. What needs to be proved is Yuen's view that warning about environmental disasters and campaigning to prevent
them has "damaging and rightward-leaning effects" that are so
severe that radicals cannot overcome them.

He offers no evidence. He devotes pages to describing the
efforts of reactionaries to misdirect concern about climate change,
but none to the efforts of radical environmentalists to counter
those forces. Earlier in his essay, he mentions that "environmental and climate justice perspectives are steadily gaining traction in
internal environmental debates," but those thirteen words are all
he has to say on the subject.

He says nothing about the historic 2010 Cochabamba
Conference, where 30,000 environmental activists from 140 countries warned that if greenhouse gas emissions are not stopped,
"the damages caused to our Mother Earth will be completely

irreversible," a statement Yuen would doubtless label "catastroph-ist." Far from succumbing to apathy or reactionary policies, the participants explicitly rejected market solutions, identified capi-talism as the cause of the crisis, and outlined a radical program to transform the global economy.

He is equally silent about the campaign against the fraudu-lent "green economy" plan adopted at the Rio+20 conference in 2012. One of the principal organizers of that opposition was La Via Campesina, the world's largest organization of peasants and farmers, which warns that the world's governments are "propagat-ing the same capitalist model that caused climate chaos and other deep social and environmental crises."

His essay contains not a word about Idle No More, or Occupy, or the indigenous-led fight against Canada's tar sands, or the anti-fracking, anti-coal, and anti-pipeline movements in many countries. By omitting them, Yuen leaves the false impression that the climate movement is helpless to resist reactionary forces.

Contrary to Yuen's title, the effort to build a movement to save the planet has not failed. The question before radicals is not what "narrative strategy" to adopt, but rather, how will we relate to the growing environmental movement? How will we support its goals while strengthening the forces that see the need for more radical solutions?

What Must Be Done?

Yuen says that building movements around rallies, marches, and other mass protests to get out the truth and to demand action worked in the 1960s, when Americans were well-off and naive, but cannot be replicated in today's "culture of atomized cynicism."

Like many who know that decade only from history books or as distant memories, Yuen foreshortens the experience: he knows about the mass protests and dissent late in the decade, but ignores the many years of educational work and slow movement building in a deeply reactionary and racist time. It is not predetermined that the campaign against climate change will take as long as those struggles, or take similar forms, but the real experience of the 1960s should at least be a warning against premature declarations of failure.

Yuen is much less explicit about what he thinks would be an effective strategy, but he cites as positive examples the efforts of some to promote "a bottom-up and egalitarian transition" by

> ever-increasing numbers of people who are voluntarily engaging in intentional communities, sustainability projects, permaculture and urban farming, communing and militant resistance to consumerism. . . . We must consider the alternative posed by the highly imaginative Italian left of the twentieth century. The explosively popular Slow Food movement was originally built on the premise that a good life can be had not through compulsive excess but through greater conviviality and a shared commonwealth.

Compare that to this list of essential tasks, prepared by Pablo Solón, a leading figure in the global climate-justice movement, who says that to reduce greenhouse gas emissions to a level that avoids catastrophe, we must

- Leave more than two-thirds of the fossil fuel reserves under the soil.
- Stop the exploitation of tar sands, shale gas, and coal.
- Support small, local, peasant, and indigenous community farming while we dismantle big agribusiness that deforests and heats the planet.
- Promote local production and consumption of products, reducing the free trade of goods that moves millions of tons of CO_2 while they travel around the world.
- Stop extractive industries from further destroying nature and contaminating our atmosphere and our land.
- Significantly increase public transport to reduce the unsustainable "car way of life."

- Reduce the emissions of warfare by promoting genuine peace and dismantling the military and war industry and infrastructure.[20]

The projects that Yuen describes are worthwhile, but unless the participants are also committed to building mass environmental campaigns, they will not be helping to achieve the vital objectives that Solón identifies. Posing local communes and slow food as alternatives to building a movement against global climate change is effectively a proposal to abandon the fight against capitalist eco-cide in favor of creating greenish enclaves while the world burns.

Bright-Siding versus Movement-Building

Whatever its merits in other contexts, it is not helpful or appropri-ate to use the word *catastrophism* as a synonym for telling the truth about the environmental dangers we face. Using the same language as right-wing climate science deniers gives the impression that the dangers are nonexistent or exaggerated. Putting accurate environ-mental warnings in the same category as apocalyptic Christian fundamentalism and century-old misreadings of Marxist eco-nomic theory leads to underestimation of the threats we face and directs efforts away from mobilizing an effective counterforce.

Yuen's argument against publicizing the scientific consensus on climate change echoes the myth that liberal politicians and jour-nalists use to justify their failure to challenge the crimes of the fossil-fuel industry. People are tired of all that doom and gloom, they say. It is time for positive messages! Or, to use Yuen's vocabu-lary, environmentalists need to end "apocalyptic rhetoric" and find better "narrative strategies." This is fundamentally an elitist position: the people cannot handle the truth, so a knowledge-able minority must sugarcoat it to make the necessary changes palatable.

David Spratt of the Australian organization Climate Code Red calls that approach "bright-siding," a reference to the bitterly satirical Monty Python song, "Always Look on the Bright Side of Life." The problem is, Spratt writes: "If you avoid including an honest assessment of climate science and impacts in your narrative, it's pretty difficult to give people a grasp about where the climate system is heading and what needs to be done to create the conditions for living in climate safety, rather than increasing and eventually catastrophic harm."[21]

Joe Romm makes the same point: "You'd think it would be pretty obvious that the public is not going to be concerned about an issue unless one explains why they should be concerned."[22]

Of course, this does not mean that we only need to explain the science. We need to propose concrete goals, as Pablo Solón has done. We need to show how the scientific consensus about climate change relates to local and national concerns such as pipelines, tar sands, fracking, and extreme weather. We need to work with everyone who is willing to confront any aspect of the crisis, from people who still have illusions about capitalism to convinced revolutionaries. Activists in the wealthy countries must be unstinting in their political and practical solidarity with the primary victims of climate change, indigenous peoples, and impoverished masses everywhere.

We need to do all of that and more. But the first step is to tell the truth about the danger we face, about its causes, and about the measures that must be taken to turn back the threat. In a time of universal deceit, telling the truth is a revolutionary act.

⚬

UPDATE: Eddie Yuen replied to this article, and I responded, in *Monthly Review*, December 2013. The exchange didn't cover any new ground, but it can be read on the *MR* website. Another exchange on catastrophism, between Canadian socialist Sam Gindin and me, was published in *Climate & Capitalism* in July and August 2014.

CHAPTER 12

Ecosocialists and the Fight for Climate Justice

SW: The UN climate talks are starting in Paris. What can we expect from these? Is there any chance there will be a significant agreement that comes out of them?
IA: My guess is that they will try to produce something that looks better than the Copenhagen accord, and there's a good chance that they'll succeed. The real issue is whether the agreement means much in practice.

All of the major countries have announced targets for emissions reductions. But as many people have pointed out, even if every single one of those targets is achieved—and we know they won't be—the temperature increase will still be way over two degrees Celsius. In addition, most of the countries that made those submissions haven't actually made concrete plans to do anything. They've announced goals, but there's no program for achieving them.

Adapted from "Climate Change and the Summit Smokescreen," an interview with Ian Angus in *Socialist Worker* (US), December 2, 2015.

It remains to be seen whether anything serious is adopted. I don't think we will get mandatory reductions. That seems extremely unlikely. And of course there's always the possibility that the whole thing could blow up, but that also seems unlikely. I think Obama wants to get something at this point in his career, so I suspect we will get good theater and something that looks okay on paper, and not much will happen as a result.

SW: Why is it so difficult for them to come up with a serious proposal that will actually make a difference to the climate situation?
IA: Government negotiators say "we'll reduce emissions," but they don't say "we will reduce our use of fossil fuels," which is what they must do to reduce emissions seriously and long term.

Fossil fuels are so fundamental to the operation of capitalism and the world today that serious reductions, if they were even tried under capitalism, would lead to a period of extraordinary economic disruption. Entire industries would have to stop functioning while they retool, and other industries would just have to disappear.

The only time we've had significant reductions in emissions in a major country was after the breakup of the Soviet Union, when the economy of Russia collapsed. And even that didn't produce the level of emissions reductions we'd like to see. So the difficulty in reaching agreement is, ultimately, that they aren't willing to reorient their economies away from fossil fuels, because fossil fuels are embedded in the way capitalism works.

SW: So how serious is the situation that we're currently facing? Every week, it seems like there's a new report about the impact of climate change, and things look worse than we thought before. Some people think it's too late already. Two degrees Celsius is already much more than the environment could take. Even if we could keep warming to one degree, we'd still be facing very, very serious problems. What's your take on that? How much time do we have to change things?
IA: It's hard to say how much time we have, because climate is so

complex. You can say that if emissions change and all other variables stay the same, then we will have some particular result in twenty years. But of course the variables don't stay the same, they change constantly. So what we're talking about is not certainties, but probabilities.

I don't like to put a time limit on it, because we're going to have to continue fighting no matter what happens. If the global temperature rises two degrees Celsius, we're going to have to fight to keep it from increasing by three degrees, and so on. But I will say that without radical economic change, the two-degree objective can't be achieved.

The scenarios used by the Intergovernmental Panel on Climate Change (IPCC) show that keeping the global temperature increase to less than two degrees by 2100 would require much greater emissions reductions than anyone is proposing in the next thirty years. And after 2050 it would require "negative emissions." That is, some technology would have to be invented that takes carbon dioxide out of the air. No such technology exists, and if it is invented, no one can say how it would function on a global scale, or whether it will be safe. It's pure fantasy, and we can't depend on fantasy.

The way things are going now, keeping the global temperature increase to two degrees is very unlikely. Without radical economic change, it's more likely that we will have a three-degree increase by the end of the century, and maybe four. That, as we know, would be catastrophic. There would be substantial parts of the earth that would be very difficult, even impossible, to survive in. So we're in a dangerous situation.

I don't see a serious commitment by any of the large nations participating in the UNFCCC [United Nations Framework Convention on Climate Change] to actually make the necessary changes. Eleven or twelve countries could actually make a difference, but none are really trying.

SW: What would radical action really look like? One of the debates in the movement has been about growth versus degrowth. Some

people argue that any kind of growth of the economy is ruled out by the danger of climate change, and that we have to move to a simpler kind of social organization, simpler technology, and so on. What do you make of that debate?

IA: The degrowth movement is mostly in Europe, mainly in France. It includes some good people doing very good analysis of the problems. The difficulty is that they focus on growth as an abstraction—it's all "bigger is bad." Instead of targeting the kind of growth that you get in a system that's based on commodity production and on capital accumulation, they often seem only to be against "more stuff."

To save the planet, we have to stop some significant things. Two really good immediate steps would be to shut down the armed forces and ban all advertising. Both of those are trillion-dollar-a-year items with huge environmental impacts. Any government that was really committed to stopping environmental destruction would take those steps. You could call that degrowth—stop doing the things that are causing the damage.

On the other hand, we are never going to build a global movement unless we recognize and accept that two-thirds of the world actually needs "more stuff." For example, we need to make access to electricity in every home a basic right. That's going to require a lot of solar panels and other equipment. There is no way around that. So focusing on reducing or stopping growth in general doesn't get us very far.

SW: You were one of the early proponents of ecosocialism. What exactly is ecosocialism? What type of contribution do you see ecosocialists making to the environmental movement?

IA: In every talk that I've ever given on ecosocialism, I've said that there is no trademark on the word. The range of opinions about what constitutes ecosocialism is very broad, just like the range of opinions about what constitutes socialism.

Think of all the different variations of socialism you've heard of, and then add all the variations of ecology to that. There are green social democrats and green anarchists and green revolutionary Marxists. There are even a few hard-core Malthusians who call themselves ecosocialists. I think they're mistaken, but that's their opinion.

Ecosocialism is three different things. First, it's a *goal*—a society in which capitalism no longer dominates and that places high priority on repairing the ecological damage that has been done and on ensuring that we don't do any more.

Second, it's a *body of ideas*. John Bellamy Foster talks about first-stage and second-stage ecosocialism. The first wave, in the 1990s, attempted to combine green political thought and Marxism. Some very important analysis resulted, but also a lot that was politically incoherent, because there are areas in which Marxism and traditional green political thought are simply not compatible.

The second wave began with two books published in 1999 and 2000: Paul Burkett's *Marx and Nature* and Foster's *Marx's Ecology*. They both, in different ways, asked, "What did Marx actually say about humanity's relationship with nature in capitalist society?" They showed that Marx said an awful lot more about ecology than most twentieth-century Marxists realized, let alone than what some green critics believed. That led to the elaboration of metabolic rift theory, which some environmentalists are now using to understand specific environmental problems.

Third, ecosocialism is a *movement*. It includes a fair range of opinion, but fundamentally it's composed of people who agree that no ecological revolution can succeed if isn't socialist, and that no socialist revolution can succeed if it isn't ecological.

Ecosocialism involves recognition that the environmental crisis is the biggest problem we face in the twenty-first century. If we don't recognize its centrality, our politics will be irrelevant.

Marx famously said that people make their own history, but not under conditions of their choosing. Changing the world in the context of impending environmental disaster is a concrete example of that. Marx didn't expect it, but that's our reality. The way we build socialism, the kind of socialism we will be able to build, will be fundamentally shaped by the state of the planet we must build it on.

SW: Given that, what strategies should we be advocating for in the broad climate justice movement?

IA: First of all, I think we have to accept that the socialist movement is not going to triumph in the immediate future. So we have to find ways to unite the broadest possible range of people, socialists or not, in a mass movement to achieve whatever is possible in this social order. We need to work with everyone who is willing to join in fighting climate change in general, and the fossil-fuel industry specifically.

That's going to include everyone who supports Naomi Klein's approach, 350.org and other radical NGOs, and also local groups, including many people who are not socialists, but who see this as the central issue in our time. Our key strategy must be to build that broad movement.

We have to accept that in such a movement, we're outnumbered. There are far more people from non-socialist backgrounds than there are of us. We won't always agree with specific actions or slogans or demands, but that's just how it's going to be. Standing on the sidelines criticizing will get us precisely nowhere: socialists must be in the movement, building it to the best of our ability.

As I said, I think the focus should be on the fossil-fuel industry. That doesn't mean we won't do other things, but we must give priority to fighting fossil fuels because that's where such a movement can actually have a substantial impact, even if we can't change the entire system yet. If we can't shut down a pipeline or prevent fracking someplace or get a university to divest itself of investments in the oil industry, how can we imagine that we're actually going to overthrow capitalism? A socialist movement that doesn't take defending human survival as a central goal isn't worthy of the name.

SW: One of the things that the movement has to figure out is how to operate in an environment dominated by NGOs, which bring some

positives to the movement, but also have a very different way of operating. Do you have any thoughts about that general question?

IA: There are NGOs, and there are NGOs.

Most of the "Big Green" NGOs have long ago given up on any serious effort to save the environment. They're fundamentally fundraising operations: they lobby, and they raise money, but that's it. If we can get people associated with those organizations to speak at rallies, to sponsor events, that's wonderful, and we should press them to do more, but we shouldn't expect them to lead the fight.

But there are smaller, more activist-oriented groups, like 350. org and the Leap Manifesto people in Canada. On the whole, they are serious and committed, but they have no tradition of, and no apparent interest in, building democratic mass movements. Despite that, we have to find ways to work with them, even if it just means agreeing to march together.

Unfortunately, a lot of folks on the left seem to think that the main thing we need to do is separate ourselves from other activists by focusing on where we disagree. Some people on the left have gotten so used to being isolated that they are uncomfortable when the walls start to break down. They feel safer in isolation.

I saw a lot of that when Naomi Klein's book *This Changes Everything* came out. It attracted a lot of attention and created openings for discussions of capitalism that we hadn't seen for a long time. Instead of treating that as a major step forward, many leftists promptly denounced it and her for not being *really* anti-capitalist.

Or take the Pope's encyclical, which was vastly superior to what we've seen from most NGOs and mainstream environmental groups. Not perfect by any means, but an opening for us to talk to religious people and involve them in the climate movement.

But when I wrote a favorable review of it for *Climate & Capitalism*, I promptly got e-mails from people saying, "Yeah, but he opposes abortion." Of course we oppose his awful views on women, but if we really want to build a broad movement, we need to learn how to work with and talk to people we disagree with.

SW: Another big piece of building the movement is the labor movement. There have been some sections of the labor movement that have taken the issue of climate change seriously, but an awful lot more who have yet to do so. What are your thoughts about how we need to win over organized labor, because that's going to be an important part of the coalition fighting on these issues.

IA: One of the difficulties, of course, is that the labor movement has focused so much of its attention for half a century on ensuring that its members have jobs, making sure that the industry they are in gets its share of government contracts, and so on. So they see themselves as having a vested interest in pipeline-building or drilling oil wells or whatever.

The fact is that workers don't want to lose their jobs. Here in Canada, we have the phenomenon of people from some of the poorest parts of the country going to work in the Alberta tar sands. After six months or a year, they can go home, to a place where there are no jobs, and buy a house or a car, or pay off their debts. Telling those people "Don't do that because you're causing greenhouse gas emissions" is just absurd. It's a guaranteed way to turn working people against the environmental movement.

Unfortunately, we see a lot of that. I've heard greens argue that we shouldn't even try to reach tar-sands workers because they're part of the colonial-settler assault on First Nations territory. Which is true. So we have to win them away from doing that, not force them into a firmer alliance with their bosses.

We need to find ways to work with the labor movement around the whole concept of a just transition. That concept has come out of the international labor movement—that the change in the economy is going to result in lost jobs, and nobody should suffer as a result. There should be jobs or full pay, free retraining, and so on.

But those things won't be won from the outside. They won't be won at all unless the unions themselves build a movement to fight for them.

CHAPTER 13

Ecosocialism: A Society of Good Ancestors

n June 1992, 172 governments, including 108 heads of state, met at the Earth Summit in Rio de Janeiro.

That meeting produced the United Nations Framework Convention on Climate Change, the first international agreement that aimed "to achieve stabilization of greenhouse gas concentrations in the atmosphere at a low enough level to prevent dangerous anthropogenic interference with the climate system." In particular, the industrialized countries promised to reduce their greenhouse gas emissions below 1990 levels.

That agreement, and the Kyoto Accord that followed it, failed miserably. For 25 years, the world's top politicians have demonstrated their gross hypocrisy and their indifference to the future of humanity and nature by giving fine speeches and making promises and then continuing with business as usual as climate change accelerates toward disaster. Global emissions today are 60% higher than in 1990.

Adapted from a talk by Ian Angus at the conference "World at a Crossroads: Fighting for Socialism in the 21st Century," in Sydney, Australia, April 2009.

But there was one exception, one head of state who spoke out strongly in Rio and called for immediate emergency action, and then returned home to support implementation of practical policies for sustainable, low-emission development. That head of state was Fidel Castro.

Fidel began his brief remarks to the plenary session of the 1992 Earth Summit with a blunt description of the crisis: "An important biological species is in danger of disappearing due to the fast and progressive destruction of its natural living conditions: humanity. We have become aware of this problem when it is almost too late to stop it."

He placed the blame for the crisis squarely on the imperialist countries, and he finished with a warning that emergency action was needed: "Tomorrow it will be too late to do what we should have done a long time ago."[1]

After the 1992 Earth Summit, only the Cubans acted on their promises and commitments.

Cuba amended its constitution to recognize the importance of "sustainable economic and social development to make human life more rational and to ensure the survival, well-being and security of present and future generations." The amended constitution obligates governmental institutions, the provincial and municipal assemblies of People's Power, to implement and enforce environmental protections. And it says that "it is the duty of citizens to contribute to the protection of the waters, atmosphere, the conservation of the soil, flora, fauna and nature's entire rich potential."[2]

The Cubans adopted low-fertilizer agriculture and encouraged urban farming to reduce the distances food has to travel. They replaced all of their incandescent light bulbs with fluorescents, and every family received an energy-efficient rice cooker. They stepped up reforestation, more than doubling the island's forested area, to 30.6 percent in 2016.

As a result of these and many other projects, in 2006 the WWF's *Living Planet Report* concluded that Cuba was the only country in the world that met its criteria for sustainable development.[3]

By contrast, the countries responsible for the great majority of greenhouse gas emissions followed one of two paths. Some gave lip service to cleaning up their acts, but in practice did little or nothing. Others denied that action was needed and so did little or nothing. As a result we are now very close to the tomorrow that Fidel spoke of, the tomorrow when it is too late.

Why Cuba?

The WWF deserves credit for honestly reporting Cuba's achievements, but it failed to address the next logical question. Why was Cuba the exception? Why could a small island republic in the Caribbean do what no other country could do?

An important factor in Cuba's actions, of course, was the 1991 collapse of the Soviet Union. However, Richard Levins, the Marxist biologist and activist who worked closely with Cuban agronomists both before and after the collapse, has pointed out that the rapid transition was only possible because a move away from the Soviet model to ecological agriculture had already begun long before the crisis.

> This discussion raged in Cuba in the 1970s and by the 1980s the ecological model had basically won although implementation was still a long process. The Special Period, that time of economic crisis after the collapse of the Soviet Union when the materials for high-tech became unavailable, allowed ecologists by conviction to recruit the ecologists by necessity. This was possible only because the ecologists by conviction had prepared the way.[4]

Armando Choy, a leader of the Cuban Revolution who headed the drive to clean up Havana Harbor, offers this clear and compelling explanation for the success of Cuba's ecological transformation:

> This is possible because our system is socialist in character and commitment, and because the revolution's top leadership acts

in the interests of the majority of humanity inhabiting planet Earth—not on behalf of narrow individual interests, or even simply Cuba's national interests.[5]

General Choy's comments reminded me of a passage in *Capital*, a paragraph that all by itself refutes the claim that is sometimes made that Marxism has nothing in common with ecology. Karl Marx wrote:

> Even an entire society, a nation, or all simultaneously existing societies taken together, are not the owners of the earth. They are simply its possessors, its beneficiaries, and have to bequeath it in an improved state to succeeding generations.[6]

I've never known any socialist organization to make this point explicitly, but Marx's words imply that one of the key objectives of socialism must be to build a society in which human beings act consciously to be *good ancestors*. In contrast to capitalism's incurable focus on short-term gains, a socialist society must think and act in harmony with the needs of our grandchildren—and of *their* grandchildren.

This is what the Cubans are doing in practice. The idea that we must act in the present to build a better world for the future has been a theme of the Cuban revolutionary movement since Fidel's inspiring 1953 speech, "History Will Absolve Me." The commitment to future generations is central to what has justly been called the greening of the Cuban Revolution. The Cuban people are committed, not just in words but in practice, to being good ancestors, not only to future Cubans but to future generations around the globe.

Why Not Capitalism?

But what about the other side of the question? Why do we not see a similar commitment from the ruling classes of the United

States, or Australia, or Canada? Why have the richest countries in the world not cut their emissions, not developed sustainable economies? Why, despite their enormous physical and scientific resources, has their performance actually gotten worse?

If you asked any of them individually, our rulers would undoubtedly say that they want their children and grandchildren to live in a stable and sustainable world. So why do their actions contradict their words? Why do they seem determined, in practice, to leave their children and grandchildren a world of poisoned air and water, a world of floods and droughts and escalating climate disasters? Why have they repeatedly sabotaged international efforts to adopt even halfhearted measures to cut greenhouse gas emissions?

When they do consider or implement responses to the climate crisis, why do they always support solutions that do not work, that cannot possibly work?

Karl Marx had a wonderful phrase for the bosses and their agents, the big shareholders and executives and top managers and the politicians they own, a phrase that explains why they invariably act against the present and future interests of humanity. These people, he said, are *personifications of capital*. Regardless of how they behave at home, or with their children, their social role is that of capital in human form.

They don't act to stop climate change because the changes needed by the people of this world are directly contrary to the needs of capital. Capital has no conscience. Capital can't be anyone's ancestor because capital has no children. Capital has only one imperative: it has to grow, as rapidly as possible. As I wrote in *Facing the Anthropocene*, there is "an insuperable conflict between Nature's Time and Capital's Time—between the cyclical Earth System processes that have developed over millions of years, and capital's need for rapid production, delivery, and profit."[7]

The only reason for using money to buy stock, launch a corporation, build a factory, or drill an oil well is to get more money back than you invested. That doesn't always happen, of course; some investments fail to produce profits, and periodically the entire system goes into free fall, wiping out jobs and livelihoods and destroying capital. But it remains true that the *potential for profit*, to make capital grow, is a defining feature of capitalism. Without it, the system would rapidly collapse.

Under capitalism, the only measure of success is how much is sold profitably every day, every week, every year. It doesn't matter that the sales include vast quantities of products that are directly harmful to both humans and nature, or that many commodities cannot be produced without spreading disease, destroying the forests that produce the oxygen we breathe, demolishing ecosystems, and treating our water, air, and soil as sewers for the disposal of industrial waste. It all contributes to profits, and thus to the growth of capital—and that's what counts.

In *Capital*, Marx wrote that from a capitalist's perspective, raw materials such as metals, minerals, coal, oil, etc., are "furnished by Nature gratis." The wealth of nature doesn't have to be paid for or replaced when it is used. It is there for the taking. If capitalists had to pay the real cost of replacing or restoring that wealth, their profits would fall drastically.

That's true not only of raw materials, but also of what are sometimes called "environmental services"—the water and air that have been absorbing capitalism's waste products for centuries. They have been treated as free sewers and free garbage dumps, "furnished by Nature gratis."

That's what the pioneering environmental economist K. William Kapp meant nearly seventy years ago when he wrote, "Capitalism must be regarded as an economy of unpaid costs." Kapp said that capitalism's claims of efficiency and productivity are "nothing more than an institutionalized cover under which it is possible for private enterprise to shift part of the costs to the shoulders of others and to practice a form of large-scale spoliation which transcends everything the early socialists had in mind when they spoke of the exploitation of man by man."[8]

In short, pollution is not an accident, and it is not a "market failure." It is the way the system works.

How large is the problem? In 1998 the World Resources Institute studied the resource inputs used by corporations in major industrial countries—water, raw materials, fuel, and so on. Then they determined what happened to those inputs. They found that "one-half to three-quarters of annual resource inputs to industrial economies are returned to the environment as wastes within a year."[9]

Similar numbers have been reported by others. We know that about a billion people live in hunger, and yet, as the head of the United Nations Environmental Program said recently, "over half of the food produced today is either lost, wasted or discarded as a result of inefficiency in the human-managed food chain."[10]

"Inefficiency" in this case means that there is no profit to be made by preventing food waste, so waste continues. In addition to exacerbating world hunger, capitalism's gross inefficiency poisons the land and water with food that is harvested but not eaten.

Capitalism combines an irresistible drive to grow with an irresistible drive to create waste and pollution. If nothing stops it, capitalism will expand both those processes infinitely.

But Earth is not infinite. The atmosphere and the oceans and the forests are very large, but ultimately they are finite, limited resources—and capitalism is now pressing against those limits. The 2006 WWF *Living Planet Report* concluded, "The Earth's regenerative capacity can no longer keep up with demand—people are turning resources into waste faster than nature can turn waste back into resources."

My only disagreement with that statement is that it places the blame on "people" as an abstract category. The devastation is caused by the global capitalist system and by the tiny class of exploiters that profits from capitalism's continued growth. The majority of people are victims, not perpetrators.

In particular, capitalist pollution has passed the physical limit of nature's ability to absorb carbon dioxide and other gases while keeping Earth's temperature steady. As a result, the world is warmer today than it has been for many thousands of years, and global temperatures are rising at unprecedented speed.

Greenhouse gas emissions are not unusual or exceptional. Pouring crap into the environment is a fundamental feature of capitalism, and it isn't going to stop so long as capitalism survives. That's why "solutions" like carbon trading have failed so badly and will continue to fail: waste and pollution and ecological destruction are built into the system's DNA.

No matter how carefully the scheme is developed, no matter how many loopholes are identified and plugged, and no matter how sincere the implementers and administrators may be, capitalism's fundamental nature will always prevail.

We saw that happen with the Kyoto Clean Development Mechanism, which allowed polluters in rich countries to avoid cutting their own emissions by investing in emission-reducing projects in the global South. A Stanford University study showed that two-thirds or more of the CDM emission reduction projects produced no reduction in pollution. The entire system was based on what one observer called "enough lies to make a sub-prime-mortgage pusher blush."[11] It was an attempt to trick the market into doing good in spite of itself, but capitalism's drive for profits wins every time.

From Ecocapitalism to Ecosocialism

One of the greatest weaknesses of the mainstream environmental movement has been its failure to recognize that capitalism is the root problem. Indeed, many of the world's Green parties, including the one in Canada where I live, are explicitly ecocapitalist, committed to maintaining the profit system.

Of course, this puts them in a contradictory position when they face the reality of capitalist ecocide.

In Canada, oil companies are engaged in what has been called the biggest environmental crime in history, mining the Alberta tar sands. If it continues, that project will ultimately destroy 141,000 square kilometers (54,000 square miles) of boreal forest, in order to produce oil by a process that generates much more greenhouse gas than normal oil production. It is destroying ecosystems, killing animals, fish and birds, and poisoning the drinking water used by indigenous peoples in that area.

It's obvious that anyone who is serious about protecting the environment and stopping emissions should demand that the tar sands be shut down. But when I proposed that at a meeting in Vancouver, a Green Party candidate in the audience called me irresponsible, because such action would violate the oil companies' contract rights. For that ecocapitalist, "capitalism" takes precedence over "eco."

But as capitalist destruction accelerates, and as capitalist politicians continue to stall or to introduce measures that only benefit the fossil-fuel companies, we can expect that the most sincere and dedicated greens will begin to question the system itself, not just its worst results.

A case in point is James Gustave Speth, former dean of the Yale University School of Forestry and Environmental Studies. He spent most of his life trying to save the environment by working inside the system. He was a senior environmental advisor to President Jimmy Carter, and later to President Bill Clinton. In the 1990s he was Administrator of the United Nations Development Program and Chair of the United Nations Development Group.

After forty years working inside the system, Speth wrote *The Bridge at the Edge of the World*, a book in which he argued that working inside the system has failed, because the system itself is the cause of environmental destruction.

> My conclusion, after much searching and considerable reluctance, is that most environmental deterioration is a result of systemic failures of the capitalism that we have today. . . . Inherent in the dynamics of capitalism is a powerful drive to earn profits, invest them, innovate, and thus grow the economy, typically at exponential rates.

That's exactly correct. No Marxist could have said it better. Nor could we improve on Speth's summary of the factors that combine to make capitalism the enemy of ecology.

> An unquestioning society-wide commitment to economic growth at almost any cost; enormous investment in technologies designed with little regard for the environment; powerful corporate interests whose overriding objective is to grow by generating profit, including profit from avoiding the environmental costs they create; markets that systematically fail to recognize environmental costs unless corrected by government; government that is subservient to corporate interests and the growth imperative; rampant consumerism spurred by a worshipping of novelty and by sophisticated advertising; economic activity so large in scale that its impacts alter the fundamental biophysical operations of the planet; all combine to deliver an ever-growing world economy that is undermining the planet's ability to sustain life.[12]

Speth is no Marxist. He still hopes that governments can reform and control capitalism, eliminating pollution. He's wrong about that, but his account of the problem is dead-on, and the fact that it comes from someone who worked for so long inside the system makes his argument against capitalism credible and powerful. The socialist movement should welcome and publicize this development, even though Speth and others like him don't yet take their ideas to the necessary socialist conclusions.

Why Ecosocialism?

That brings me to a question I've been asked many times: "Why *eco*socialism?" Why not just say "socialism"? Marx and Engels were deeply concerned about humanity's relationship to nature, and what today we would call ecological ideas are deeply embedded in their writings. So why do we need a new word?

All that is true. But it is also true that during the twentieth century socialists forgot or ignored that tradition, supporting (and in some cases implementing) approaches to economic growth and development that were grossly harmful to the environment.

In 2009, Oswaldo Martinez, president of the Economic Affairs Commission of Cuba's National Assembly, addressed the issue bluntly:

> The socialism practiced by the countries of the Socialist Camp replicated the development model of capitalism, in the sense that socialism was conceived as a quantitative result of growth in productive forces. It thus established a purely quantitative competition with capitalism, and development consisted in achieving this without taking into account that the capitalist model of development is . . . inconceivable for humanity as a whole.
>
> The planet would not survive. It is impossible to replicate the model of one car for each family, the model of the idyllic North American society, Hollywood etc.—absolutely impossible, and this cannot be the reality for the 250 million inhabitants of the United States, with a huge rearguard of poverty in the rest of the world.
>
> It is therefore necessary to come up with another model of development that is compatible with the environment and has a much more collective way of functioning.[13]

In my view, one good reason for using the word ecosocialism is to signal a clear break with the practices that Martinez describes, practices that were called socialist for seventy years. It is a way of saying that we aim not to create a society based on having more things, but on living better. Not quantitative growth, but qualitative change.

Another reason, just as important, is to signal loud and clear that we view the global ecological crisis not as just another stick to bash capitalism with, but as the principal problem facing humanity in this century.

Close the Circle

In a short memoir published in 2008, Richard Levins explained the importance in his life of Marx's famous Eleventh Thesis on Feuerbach: *The philosophers have only interpreted the world, in various ways; the point is to change it.*

Levins described how twentieth-century developmental economists believed that simply expanding economic activity would soon eliminate poverty, and how biologists believed that rising living standards would eliminate infectious diseases. Both groups were wrong, as the continuing prevalence of hunger and the late-century upsurge of malaria, cholera, AIDS, and other deadly plagues shows. Capitalism's inability to end poverty and disease, Levins wrote, is part of a planetary crisis that requires revolutionary solutions:

> The resurgence of infectious disease is but one manifestation of a more general crisis: the eco-social distress syndrome—the pervasive multilevel crisis of dysfunctional relations within our species and between it and the rest of nature. It includes in one network of actions and reactions patterns of disease, relations of production and reproduction, demography, our depletion and wanton destruction of natural resources, changing land use and settlement, and planetary climate change. It is more profound than previous crises, reaching higher into the atmosphere, deeper into the earth, more widespread in space, and more long lasting, penetrating more corners of our lives. It is both a generic crisis of the human species and a specific crisis of world capitalism. Therefore it is a primary concern of both my science and my politics. . . .
>
> These interests inform my political work: within the left, my task has been to argue that our relations with the rest of nature cannot be separated from a global struggle for human liberation, and within the ecology movement my task has been to challenge the "harmony of nature" idealism of early

environmentalism and to insist on identifying the social relations that led to the present dysfunction. At the same time
my politics have determined my scientific ethics. I believe
that all theories are wrong that promote, justify, or tolerate
injustice.[14]

There, in a few words, is the case for ecosocialism, a political
and social movement that is founded on the intersection of science
and socialism, in the firm conviction, to quote another essay by
Levins, that "socialism cannot succeed without committing to an
ecological pathway."[15]

Evo Morales, president of Bolivia and first indigenous head of
state in Latin America, may never have used the word ecosocialism, but he has powerfully defined the problem, named the villain,
and posed the alternative:

> Competition and the thirst for profit without limits of the
> capitalist system are destroying the planet. Under Capitalism
> we are not human beings but consumers. Under Capitalism,
> Mother Earth does not exist, instead there are raw materials.
> Capitalism is the source of the asymmetries and imbalances in
> the world. It generates luxury, ostentation and waste for a few,
> while millions in the world die from hunger in the world.
>
> In the hands of capitalism everything becomes a commod-
> ity: the water, the soil, the human genome, ancestral cultures,
> justice, ethics, death . . . and life itself. Everything, absolutely
> everything, can be bought and sold under capitalism. And
> even "climate change" itself has become a business.
>
> "Climate change" has placed all humankind before a great
> choice: to continue in the ways of capitalism and death, or to
> start down the path of harmony with nature and respect for
> life.[16]

Nearly fifty years ago Barry Commoner, who embodied the
intersection of science and socialism in his life's work, warned that

humanity's very survival is threatened by the fast-growing rift in the metabolic relationship between human society and the rest of nature. "The present system of production is self-destructive," he wrote. "The present course of human civilization is suicidal."[17]

> Human beings have broken out of the circle of life, driven not by biological need, but by the social organization which they have devised to "conquer" nature: means of gaining wealth that are governed by requirements conflicting with those which govern nature. The end result is the environmental crisis, a crisis of survival. Once more, to survive, we must close the circle. We must learn how to restore to nature the wealth that we borrow from it.[18]

The causes of the global crisis are well-known, and the deadly implications of continuing business as usual are clear. The only alternative to capitalism's catastrophic disruption of the Earth System is ecosocialism, a society of good ancestors that will give top priority to preserving our planet's life support systems, to repairing the damage that has already been done, and to ending the gross inequality and oppression that define our world today.

Acknowledgments

Most of the articles in this book were written for the online ecosocialist journal *Climate & Capitalism*. In ten years we have published some 2,500 articles, statements, reviews, interviews, and talks from around the world. I cannot imagine a more satisfying way to spend my time than working with and learning from the ecosocialist activist writers on six continents who have contributed articles, ideas, criticisms, and compliments. Thank you all!

John Bellamy Foster and Lis Angus carefully read and commented on several drafts of *A Redder Shade of Green*. Their many suggestions strengthened every chapter.

This project began in September 2016, when I informally asked Monthly Review Press if they might be interested. Not only did they quickly say yes, but they proposed very tight deadlines so that it could be included in their 2017 spring catalog. Fortunately, I was working with the remarkable team at MRP—Michael Yates, Susie Day, and Martin Paddio—and with copy editors Jess Shulman and Erin Clermont, so the deadlines were met and you have the result in your hands. I hope you find it a useful addition to your red-and-green bookshelf.

Notes

References to the *Marx-Engels Collected Works* (New York: International Publishers, 1975–2004) are abbreviated *MECW*, followed by volume and page numbers.

INTRODUCTION: ESSENTIAL DEBATES AT THE INTERSECTIONS OF
SCIENCE AND SOCIALISM

1.   Karl Marx, *Economic and Philosophical Manuscripts of 1844* (New York: International Publishers, 1964), 143.

2.   John Bellamy Foster, "Why Ecological Revolution?," *Monthly Review* (January 2010): 15.

3.   *Climate & Capitalism* blog: http://climateandcapitalism.com/. *Climate & Capitalism* Facebook page: https://www.facebook.com/groups/121827024509492/.

PART ONE: NATURAL SCIENCE AND THE MAKING
OF SCIENTIFIC SOCIALISM

1.   Helena Sheehan, "Marxism and Science Studies: A Sweep through the Decades," *International Studies in the Philosophy of Science* (July 2007): 197.

1. MARX AND ENGELS AND THE RED CHEMIST

1.   Paul Thomas, *Marxism and Scientific Socialism: From Engels to Althusser* (New York: Routledge, 2008), 11, 45.

2. Terence Ball, "Marx and Darwin: A Reconsideration," *Political Theory* (November 1979): 478.

3. Terence Ball, "Marxian Science and Positivist Politics," in Terence Ball and James Farr, *After Marx* (Cambridge: Cambridge University Press, 1984), 235.

4. Kohei Saito, "Why Ecosocialism Needs Marx," *Monthly Review* (November 2016): 60.

5. Kohei Saito, "Marx's Ecological Notebooks," *Monthly Review* (February 2016): 25–26.

6. An important exception to the general neglect of Schorlemmer's contributions to Marxism is in John L. Stanley and Ernest Zimmerman, "On the Alleged Differences Between Marx and Engels," in John L. Stanley, *Mainlining Marx* (New Brunswick, NJ: Transaction Publishers, 2002), 31–61. W. O. Henderson's *The Life of Friedrich Engels* (London: Frank Cass, 1976) includes a short account of Schorlemmer's life, but it is silent on his contributions to Marxist theory.

7. W. O. Henderson, *The Life of Friedrich Engels*, vol. 1 (London: Frank Cass, 1976), 262–71.

8. In his 1892 obituary tribute to Schorlemmer, Engels said they met at the beginning of the 1860s, but his earliest mention of the chemist is in a letter to Marx dated March 6, 1865, in *MECW*, 42:117.

9. Engels to Marx, May 10, 1868, in *MECW*, 43:33.

10. Quoted in Rachel Holmes, *Eleanor Marx: A Life* (London: Bloomsbury, 2014), 359.

11. Frederick Engels, "Karl Marx's Funeral," in *MECW*, 24:471. This translation is from Philip Foner, ed., *Karl Marx Remembered* (San Francisco: Synthesis Publications, 1983), 43.

12. Engels to Eduard Bernstein, February 27, 1883, in *MECW*, 46:446.

13. *MECW*, 27:305.

14. Engels to Marx, March 6, 1865, in *MECW*, 42:117.

15. Engels, "Carl Schorlemmer" (1892), in *MECW*, 27:305.

16. Theodor Benfey and Tony Travis, "Carl Schorlemmer: The Red Chemist," *Chemistry and Industry*, June 15, 1992: 441–44.

17. Marx to Engels, February 13, 1866, in *MECW*, 42:227.

18. John Bellamy Foster, *Marx's Ecology: Materialism and Nature* (New York: Monthly Review Press, 2000).

19. Marx to Engels, December 7, 1867, in *MECW*, 42:495.

20. Marx to Engels, January 3, 1868, in *MECW*, 42:507–8.

21. Marx to Engels, February 4, 1868, in *MECW*, 42:536.

22. The letter to Schorlemmer has not survived, but Marx mentioned his intention in Marx to Engels, February 11, 1870, in *MECW*, 43:426. At that time gun cotton was too unstable for practical use. A safe version was patented in 1889.

23. Engels, "Carl Schorlemmer," in *MECW*, 27:305.

24. Engels to Marx, May 30, 1873, in *MECW* 44:500–501 (emphasis in original).

25. Marx to Engels, May 31, 1873, in *MECW* 44:500 and 503.

26. *MECW*, 44:500–501 includes Schorlemmer's marginal comments as footnotes.

27. Thomas, *Marxism and Scientific Socialism*, 46.

28. Stanley, *Mainlining Marx*, 51.

29. *MECW*, 47:602n.

30. Engels to Paul Lafargue, May 19, 1885, in *MECW*, 47:289.

31. Carl Schorlemmer, *The Rise and Development of Organic Chemistry*, rev. ed. (London: Macmillan, 1894), 142. This edition was edited by Schorlemmer but not published until after his death. The translation of the *Anti-Dühring* sentence in *MECW*, 25:118, is slightly different.

32. Engels to Ludwig Schorlemmer, December 1, 1892, in *MECW*, 50:45.

33. Liz Else, "Slavoj Žižek: Wake Up and Smell the Apocalypse," *New Scientist*, August 25, 2010; Jason W. Moore, "Name the System! Anthropocenes & the Capitalocene Alternative," October 9, 2016, https://jasonwmoore.wordpress.com/.

34. Marx and Engels, "The German Ideology" (1846), in *MECW*, 5:28n.

35. Marx, "Theses on Feuerbach," in *MECW*, 5:3.

2. MARX AND ENGELS AND DARWIN

1. Engels to Marx, December 11 or 12, 1859, in *MECW*, 40:551.

2. Karl Marx and Frederick Engels, *Selected Correspondence 1846–1895* (New York: International Publishers, 1975), 126. In *MECW*, 41:232, this passage is translated as "the book which, in the field of natural history, provides the basis for our view."

3. Marx to Ferdinand Lassalle, January 16, 1861, in *MECW*, 41:246–47.

4. *Reminiscences of Marx and Engels* (Moscow: Foreign Languages Publishing House, n.d.), 106.

5. Karl Marx, *Capital*, vol. 1, in *MECW*, 35:346n.

6. John Bellamy Foster, *Marx's Ecology: Materialism and Nature* (New York: Monthly Review Press, 2000), 207. Darwin sent a polite thank-you note, but did not read the book.

7.   Engels, "Karl Marx's Funeral," in *MECW*, 24:467.

8.   Allan Megill, *Karl Marx: The Burden of Reason* (Lanham, MD: Rowman & Littlefield, 2002), 54.

9.   Naomi Beck, "The *Origin* and Political Thought," in *The Cambridge Companion to the Origin of Species*, ed. Michael Ruse and Robert J. Richards (Cambridge: Cambridge University Press, 2000), 313, 310.

10. D. A. Stack, "The First Darwinian Left: Radical and Socialist Responses to Darwin, 1859–1914," *History of Political Thought* (Winter 2000): 683–84. Stack repeated most of these comments, word for word, in *The First Darwinian Left: Socialism and Darwinism, 1859–1914* (Cheltenham: New Clarion Press, 2003), 4.

11. The rigid class hierarchy on naval ships meant that a captain could not socialize with other officers or the crew, but the navy permitted captains to bring appropriate passengers at their own expense. See Stephen Jay Gould, "Darwin's Sea Change," in *Ever Since Darwin* (New York: W. W. Norton, 1992), 28–33.

12. Charles Darwin, "An Historical Sketch of the Progress of Opinion on the Origin of Species," in the third and subsequent editions of *On the Origin of Species* (London: Penguin Books, 1985 [1861]), 53.

13. William Whewell, *History of the Inductive Sciences*, vol. 2 (London: John W. Parker, 1837), 563–65.

14. Herschel's letter to Charles Lyell was published in Charles Babbage, *The Ninth Bridgewater Treatise* (London: John Murray, 1838), 225–36. Darwin quoted it in the first paragraph of *On the Origin of Species.*

15. Francis Darwin, ed., The *Autobiography of Charles Darwin* (Fairfield, IA: 1st World Library, 2003), 16.

16. Adrian Desmond, *The Politics of Evolution: Morphology, Medicine and Reform in Radical London* (Chicago: University of Chicago Press, 1989), 4.

17. Robert Chambers, *Vestiges of the Natural History of Creation and Other Evolutionary Writings* (Chicago: University of Chicago Press, 1994 [1844]), 234.

18. James A. Secord, *Victorian Sensation: The Extraordinary Publication, Reception, and Secret Authorship of Vestiges of the Natural History of Creation* (Chicago: University of Chicago Press, 2000), 526.

19. Stephen Jay Gould, *Time's Arrow, Time's Cycle: Myth and Metaphor in the Discovery of Geological Time* (Cambridge, MA: Harvard University Press, 1987), 146–47.

20. Charles Darwin, *The Origin of Species* (London: Penguin Books,

1985 [1859]), 435. Darwin did not use the word "evolution" in *Origin*. At that time the word implied the unfolding of characteristics already present in the organism—the evolution of an embryo into an animal, for example. That concept was alien to Darwin's theory.

21. Foster, *Marx's Ecology*, 179.

22. Marx, "Afterword to the Second German Edition of *Capital*, Vol. 1," in *MECW*, 35:19.

23. Darwin, *Origin*, 108.

24. Thomas Huxley, "Criticisms of On the Origin Of Species," [1864] in *Lay Sermons, Addresses and Reviews* (London: Macmillan, 1888), 303.

25. Marx, "Economic and Philosophic Manuscripts of 1844," in *MECW*, 3:303–4.

26. Paul Heyer, *Nature, Human Nature, and Society: Marx, Darwin, Biology and the Human Sciences* (Westport CT: Greenwood Press, 1982), 49.

27. Engels, "Socialism: Utopian and Scientific," in *MECW*, 24:301.

28. Stack, *First Darwinian Left*, 684.

29. Stack discusses this in his essay, which makes his cynical dismissal of Engels's words at Marx's graveside hard to understand.

30. Quoted in Richard Hofstadter, *Social Darwinism in American Thought* (Boston: Beacon Press, 1993 [1944]), 45.

31. Engels to Pyotr Lavrov, November 12, 1875, in *MECW*, 45:107–8. This passage has been cited by historians as a criticism of Darwin's theory, but the context clearly shows that he is referring to Social Darwinism.

32. Marx and Engels, "The German Ideology," in *MECW*, 5:31.

33. Engels, "Dialectics of Nature, Notes and Fragments," in *MECW*, 25:584.

34. Ibid., 585.

35. Ernst Mayr, *One Long Argument: Charles Darwin and the Genesis of Modern Evolutionary Thought* (Cambridge, MA: Harvard University Press,1991), 50.

36. Heyer, *Nature, Human Nature, and Society*, 27.

PART TWO: RESPONDING TO THE ANTHROPOCENE

1. Will Steffen, Paul J. Crutzen, and John R. McNeill, "The Anthropocene: Are Humans Now Overwhelming the Great Forces of Nature?," *Ambio* (December 2007): 619.

3. A CHALLENGE THAT SOCIALISTS CANNOT IGNORE

1. Paul J. Crutzen and Will Steffen, "How Long Have We Been in the Anthropocene Era? An Editorial Comment," *Climatic Change* 61 (2003): 253.
2. Paul Crutzen, "Geology of Mankind," *Nature* (January 3, 2002): 23.
3. Will Steffen et al., "The Anthropocene: From Global Change to Planetary Stewardship," *Ambio* (November 2011): 746.

4. ECOMODERNISTS AND THE "GOOD ANTHROPOCENE"

1. "Notes from the Editors," *Monthly Review* (June 2014): 64.
2. Richard York and Eugene A. Rosa, "Key Challenges to Ecological Modernization Theory," *Organization & Environment* (September 2003): 274.
3. "Statement on 'Climate Pragmatism' from BTI Founders Michael Shellenberger and Ted Nordhaus," Breakthrough Institute, July 27, 2011, http://thebreakthrough.org/.
4. Clive Hamilton, "The New Environmentalism Will Lead Us to Disaster," *Scientific American* Forum, June 19, 2014, https://www.scientificamerican.com/.
5. Michael Shellenberger and Ted Nordhaus, *The Death of Environmentalism: Global Warming Politics in a Post-Environmental World* (Oakland, CA: Breakthrough Institute, 2004), 10.
6. Bill Blackwater, "The Denialism of Progressive Environmentalists," *Monthly Review* (June 2012): 21; John Bellamy Foster, "The Planetary Rift and the New Human Exemptionalism," *Organization & Environment* (September 2012): 213.
7. Paul J. Crutzen and Will Steffen, "How Long Have We Been in the Anthropocene Era? An Editorial Comment," *Climatic Change* 61 (2003): 253.
8. Will Steffen et al., "The Anthropocene: From Global Change to Planetary Stewardship," *Ambio* (November 2011): 2.
9. Will Steffen, Paul J. Crutzen, and John R. McNeill, "The Anthropocene: Are Humans Now Overwhelming the Great Forces of Nature?," *Ambio* (December 2007): 619.
10. John Bellamy Foster, Brett Clark, and Richard York, *The Ecological Rift: Capitalism's War on the Earth* (New York: Monthly Review Press, 2010), 18.
11. Editorial, "The Human Epoch," *Nature* (May 19, 2011): 254.
12. Ted Nordhaus and Michael Shellenberger, introduction to *Love Your Monsters: Post-Environmentalism and the Anthropocene*, ed. Ted

Nordhaus and Michael Shellenberger (Oakland, CA: Breakthrough Institute, 2011), Kindle edition, loc. 61.

13.   Ibid., loc. 208.

14.   Ibid., loc. 138.

15.   Clive Hamilton and Jacques Grinevald, "Was the Anthropocene Anticipated?," *The Anthropocene Review* (April 2015): 8.

16.   "Erle Ellis, Associate Professor, University of Maryland, Baltimore County," Breakthrough Institute, n.d., http://www.thebreakthrough. org/people/profile/erle-ellis.

17.   Jan Zalasiewicz et al., "When Did the Anthropocene Begin? A Mid-Twentieth-Century Boundary Level Is Stratigraphically Optimal," *Quaternary International* (October 5, 2015): 4.

18.   Quoted in Richard Monastersky, "Anthropocene: The Human Age," *Nature* (March 11, 2015): 147.

19.   William F. Ruddiman et al., "Defining the Epoch We Live In," *Science* (April 3, 2015): 39.

20.   Ted Nordhaus, Michael Shellenberger et al., "An Ecomodernist Manifesto," April 2015, http://www.ecomodernism.org/.

21.   Blackwater, "The Denialism of Progressive Environmentalists," 11.

22.   Ian Angus, "Another Attack on Anthropocene Science," *Climate & Capitalism*, January 24, 2017.

5. THE MOST DANGEROUS ENVIRONMENTALIST CONCEPT?

In 2014, Jason W. Moore posted two essays, "The Capitalocene, Part I," and "The Capitalocene, Part II," on his website, http://www.jasonwmoore.com/. In 2016, he uploaded extensively revised versions of each to Academia.edu, without explaining what he had changed or why. To avoid confusion, I have added the appropriate year, 2014 or 2016, after each title.

1.   Jason W. Moore, *Capitalism in the Web of Life: Ecology and the Accumulation of Capital* (London: Verso, 2015).

2.   Jason W. Moore, "The Capitalocene, Part II [2016]," 1; "The Capitalocene, Part I [2016]," 5; "Putting Nature to Work," in Cecilia Wee, Janneke Schönenbach, and Olaf Arndt, eds., *Supramarkt* (Gothenburg, Sweden: Irene Books, 2015), 13.

3.   Moore, "Putting Nature to Work," 7. *Oikeios* is an ancient Greek adjective meaning belonging to one's household. Theophrastus, a student of Aristotle, used *oikeios topos* to mean the "favorable place" where a plant species will flourish. It is not clear why Moore uses it here.

4. Moore, "The Rise of Cheap Nature," in *Anthropocene or Capitalocene? Nature, History and the Crisis of Capitalism,* ed. Jason W. Moore (Oakland, CA: PM Press, 2016), 83 and 89; Moore, *Capitalism in the Web of Life,* 169n; Moore, "The Capitalocene, Part I [2014]," 3.

5. Moore, "Putting Nature to Work," 19.

6. Moore, "The Capitalocene, Part I [2014]," 1.

7. Moore, "The Rise of Cheap Nature," 84.

8. Moore, *Capitalism in the Web of Life,* 176 and 180.

9. Moore, "The Rise of Cheap Nature," 81.

10. He does refer to the *title* of a 2007 paper by Will Steffen, Paul Crutzen, and John McNeill, "The Anthropocene: Are Humans Now Overwhelming the Great Forces of Nature?" (*Ambio,* December 2007: 614–62). But that title does not say "all humans," and the paper itself distinguishes between the impacts of industrial, agrarian, and hunting-gathering societies.

11. Ian Angus, *Facing the Anthropocene* (New York: Monthly Review Press, 2016), 224–30.

12. Will Steffen et al., "Planetary Boundaries: Guiding Human Development on a Changing Planet," *Science* (February 13, 2015): 9.

13. Moore, *Capitalism in the Web of Life,* 170. He gives exactly the same definition in *Anthropocene or Capitalocene?*, 81; and in "The Capitalocene, Part I [2014]," 2. None of those cite a source for the definition.

14. Moore, "The Capitalocene, Part II [2014]," 4.

15. Moore, *Capitalism in the Web of Life,* 2 and 4.

16. Ibid., 2, 5.

17. Moore, introduction to *Anthropocene or Capitalocene?*, 3.

18. Moore, "The Rise of Cheap Nature," 84.

19. Frank Oldfield and Will Steffen, "The Earth System," in Will Steffen et al., *Global Change and the Earth System* (Berlin: Springer, 2004), 7.

20. Steffen, Crutzen, and McNeill, "The Anthropocene: Are Humans Now Overwhelming the Great Forces of Nature?," 615.

21. Will Steffen et al., "The Trajectory of the Anthropocene: The Great Acceleration," *Anthropocene Review* 2/1 (2015): 93. (Emphasis added.)

22. Moore, *Capitalism in the Web of Life,* 176.

23. Ibid., 171.

24. Marxist geographer Judith Watson has questioned Moore's account

of early modern deforestation. I'm not qualified to judge, but it seems that the issue is not as clear-cut (pun intended) as Moore suggests. See "Book Review Roundtable: Disentangling Capital's Web," *Capitalism Nature Socialism* 27/2 (2016): 103–21.

25. Moore, "The Capitalocene, Part 1[2014]," 2 and 1.
26. Steffen et al., "Trajectory of the Anthropocene," 93.
27. Clive Hamilton, *Defiant Earth: The Fate of Humans in the Anthropocene* (Cambridge: Polity Press, 2017), 9–10. (Emphasis in original.)
28. Ibid., 30.
29. Jeremy Davies, *The Birth of the Anthropocene* (Oakland: University of California Press, 2016), 55.
30. Frederick Engels, *Anti-Dühring: Herr Eugen Dühring's Revolution in Science* (Moscow: Progress Publishers, 1969 [1878]), 10–11.
31. Andreas Malm, *Fossil Capital: The Rise of Steam Power and the Roots of Global Warming* (London: Verso, 2016), 391.
32. Anthony Galluzzo's Facebook page, https://www.facebook.com/galluzzoanthony?fref=ts.
33. Moore, "Rise of Cheap Nature," 81.
34. Moore, "The Capitalocene, Part I [2014]," 5; Moore, introduction to *Anthropocene or Capitalocene?*, 6.
35. Moore, "The Capitalocene, Part II [2016]," 1–4.
36. Jason W. Moore and Kamil Ahsan, "Capitalism in the Web of Life: An Interview with Jason W. Moore," *Viewpoint Magazine*, September 28, 2015, https://viewpointmag.com/2015/09/28/capitalism-in-the-web-of-life-an-interview-with-jason-moore/.
37. Moore, *Anthropocene or Capitalocene?*, back cover. The promotional text is unsigned, but it is written in Moore's distinctive style and is unlikely to have been used without his approval.
38. Jeremy Davies, "A Change of Cene," *Made Ground*, part 1, August 28, 2016, and part 2, September 11, 2016, https://madeground.com/. Davies's demolition of *Anthropocene or Capitalocene?* is particularly telling because he thought Moore's previous book, *Capitalism in the Web of Life*, was "a dazzling work of ecological theory." Obviously he did not approach the second book with negative preconceptions.
39. See Davies's review for a summary and critique of Hartley's five points.
40. Elmar Altvater, "The Capitalocene, or, Geoengineering against Capitalism's Planetary Boundaries," in Moore, *Capitalocene or Anthropocene*, 145. Jason W. Moore, "Technics and Historical

Nature: Praxis of the Capitalist World-Ecology," March 28, 2014, https://jasonwmoore.wordpress.com.

41. Davies, "A Change of Cene," part 1.

42. Pierre Bourdieu, *Science of Science and Reflexivity*, trans. Richard Nice (Chicago: University of Chicago Press, 2004), 8, 26–27.

43. Jason W. Moore, "Name the System! Anthropocenes & the Capitalocene Alternative," October 9, 2016, https://jasonwmoore. wordpress.com.

44. The 2014 version of "The Capitalocene, Part II" can be downloaded at http://www.jasonwmoore.com/Essays.html. The 2016 version is available at https://www.academia.edu/.

45. Jason W. Moore, "The Rise and Fall of Cheap Nature," posted to YouTube by Socialist Project, December 17, 2016, https://www.you-tube.com/watch?v=TjKjZBETBOI.

46. Moore, "Name the System!"

47. Will Steffen et al., "Stratigraphic and Earth System Approaches to Defining the Anthropocene," *Earth's Future* (August 12, 2016): 334. (Emphasis added.)

48. Moore, "Name the System!"

49. Ibid.

50. Massimo Pigliucci, *Nonsense on Stilts: How to Tell Science from Bunk* (Chicago: University of Chicago Press, 2010), 262.

51. A Google search for "academic branding" turns up several dozen coaches and consultants. A typical promise: "Your academic brand conveys who you are as a thought leader and how you stand out from colleagues in your field."

52. Bourdieu, *Science of Science*, 8.

53. McKenzie Wark, *Molecular Red: Theory for the Anthropocene* (London: Verso, 2015), 120.

54. Marx to Schweitzer, October 13, 1868, in *MECW*, 43:133.

55. Moore, "Name the System!" (italics in original).

56. Sy Taffel, "Mapping the Anthropocene," in *Ecological Entanglements in the Anthropocene*, ed. Nicholas Holm and Sy Taffel (New York: Lexington Books, 2017), 230–31.

57. Dipesh Chakrabarty, "The Climate of History: Four Theses," *Critical Inquiry* (Winter 2009): 219.

PART THREE: NUMBERS ARE NOT ENOUGH

1. Stephen Jay Gould, *The Mismeasure of Man* (New York: W. W. Norton, 1996), 106.

6. TUNNEL VISION AT THE ROYAL SOCIETY

1. Murray Bookchin, *Which Way for the Ecology Movement?* (Oakland, CA: AK Press, 1994), 30.
2. Susan Greenhalgh, "The Social Construction of Population Science: An Intellectual, Institutional, and Political History of 20th Century Demography," *Comparative Studies in Society and History* (January 1996): 48 and 52.
3. Frances Moore Lappé and Rachel Schurman, *Taking Population Seriously* (Oakland, CA: Food First Books, 1990), 30 and 33.
4. Nicholas Hildyard, *Too Many for What? The Social Generation of Food "Scarcity" and "Overpopulation"* (London: Corner House, 1996), http://www.thecornerhouse.org.uk/resource/too-many-what.
5. Barry Commoner, *The Closing Circle: Nature, Man, and Technology* (New York: Knopf, 1971), 247.
6. Lappé and Schurman, *Taking Population Seriously*, 33.
7. *Oxford English Reference Dictionary*, 20th ed., s.v. "demography."
8. See comments section under Robert Kunzig, "The Population Monster Knocking at our Door," *National Geographic Voices*, October 31, 2011, http://voices.nationalgeographic.com/2011/10/31/the-population-monster-knocking-at-our-door/.
9. Sarah Jane Keller, "Q&A: Stanford's Paul Ehrlich Fears the Worst for a Planet with 7 Billion Residents," *Stanford Report*, October 26, 2011, http://news.stanford.edu/news/2011/october/qanda-paul-ehrlich-102611.html.
10. Robert Engelman, "The World at 7 Billion: Can We Stop Growing Now?" *Yale Environment 360*, July 18, 2011, http://e360.yale.edu/.
11. Stephen Pacala, "Equitable Solutions to the Carbon and Climate Problem," lecture delivered at Stanford University's Energy Seminar, November 5, 2008, available at https://www.youtube.com/watch?v=2X2u7-R3Wrc.

7. THE RETURN OF THE POPULATION BOMBERS

1. Ian Angus and Simon Butler, *Too Many People? Population, Immigration and the Environmental Crisis* (Chicago: Haymarket Books, 2011).
2. *People and the Planet* (London: Royal Society Science Policy Centre, 2012), https://royalsociety.org/; "IAP Statement on Population and Consumption," June 2012, http://www.interacademies.net/.
3. Alan Weisman, *The World Without Us* (Toronto: HarperCollins, 2007), 272.

4. Chris Hedges, "We are breeding ourselves to extinction," *Truthdig*, March 8, 2009, http://www.truthdig.com/.

5. David Attenborough, quoted in rotating header image at http://www.popoffsets.org/.

6. William E. Rees, "Globalization, Trade and Migration: Undermining Sustainability," *Ecological Economics* 59 (2006): 223.

7. Quoted in Joseph A. Califano, *The Triumph & Tragedy of Lyndon Johnson: The White House Years* (New York: Simon & Schuster, 2015), 149.

8. Quoted in John Bellamy Foster and Brett Clark, "Rachel Carson's Ecological Critique," *Monthly Review* (February 2008): 1.

9. Lewis Herber [Murray Bookchin], *Our Synthetic Environment* (New York: Knopf, 1962), 11, https://libcom.org/files/Bookchin%20M.%20 Our%20Synthetic%20Environment.pdf.

10. Barry Commoner, *Science & Survival* (New York: Viking, 1966), 124.

11. Paul Ehrlich (and Anne Ehrlich, uncredited), *The Population Bomb* (New York: Ballantine Books, 1968), 66–67.

12. Quoted in Michael Egan, *Barry Commoner and the Science of Survival* (Cambridge, MA: MIT Press, 2007), 130.

13. Murray Bookchin, *Re-Enchanting Humanity* (London: Cassell, 1995), 60.

14. Lourdes Arizpe and Margarita Velásquez, "The Social Dimensions of Population," in *Population and the Environment: Rethinking the Debate,* ed. Lourdes Arizpe, M. Priscilla Stone, and David C. Major, (Boulder, CO: Westview Press, 1994), 18.

15. Population Matters, "Well Done, The Royal Society," April 26, 2012, https://www.populationmatters.org/royal-society/.

16. Alan Roberts, *The Self-Managing Environment* (London: Allison and Busby, 1979), 7.

17. William Hazlitt, "On the Application of Mr Malthus' Principle to the Poor Laws" (1819), http://www.blupete.com/Literature/Essays/ Hazlitt/Political/MalthusIV.htm.

8. DID NASA PREDICT CIVILIZATION'S COLLAPSE?

1. Nafeez Ahmed, "NASA-Funded Study: Industrial Civilisation Headed for 'Irreversible Collapse'?," *The Guardian*, March 14, 2014, https://www.theguardian.com/.

2. "NASA Statement on Sustainability Study," March 20, 2014, http://www.nasa.gov/.

3.   Patricia McAnany and Norman Yoffee, eds., *Questioning Collapse: Human Resilience, Ecological Vulnerability, and the Aftermath of Empire* (Cambridge: Cambridge University Press, 2010). For a critique of Catton's book *Overshoot*, see George Bradford, "How Deep Is Deep Ecology" (New York: Times Change Press, 1989), https://theanarchistlibrary.org/. For a general critique of populationism and carrying capacity theory applied to human society, see chapters 3 through 6 of Ian Angus and Simon Butler, *Too Many People? Population, Immigration and the Environmental Crisis* (Chicago: Haymarket Books, 2011).

4.   Daniel Botkin, *Discordant Harmonies: A New Ecology for the Twenty-First Century* (New York: Oxford University Press, 1990), 46.

5.   Joel E. Cohen, *How Many People Can the Earth Support?* (New York: W. W. Norton, 1995), 232 and 262.

6.   Quoted in Ian Angus, "Malthus with a Computer," *Climate & Capitalism*, November 10, 2010.

7.   Joseph A. Tainter, "Archaeology of Overshoot and Collapse," *Annual Review of Archaeology* (2006): 71. (Emphasis added.)

8.   Paul Kingsnorth, "The Quants and the Poets," April 21, 2011, http://paulkingsnorth.net/2011/04/21/the-quants-and-the-poets/.

9.   Frederick Engels to C. Schmidt, August 5, 1890, in *MECW*, 49:8.

PART FOUR: SAVING SPECIES, SAVING OCEANS

1.   Rachel Carson, *Silent Spring* (New York: Houghton Mifflin, 2002 [1962]), 99.

9. THIRD WORLD FARMING AND BIODIVERSITY

1.   Ivette Perfecto, John Vandermeer, and Angus Wright, *Nature's Matrix: Linking Agriculture, Conservation and Food Sovereignty* (London: Earthscan, 2009).

10. MORE PLASTIC THAN FISH

1.   Barry Commoner, *The Closing Circle: Nature, Man and Technology* (New York: Alfred E. Knopf, 1971), 177.

2.   Ibid., 39.

3.   Ibid., 11–12.

4.   Tarique Zaman, "The Prevalence and Environmental Impact of Single Use Plastic Products," http://www.cwru.edu/med/epidbio/mphp439/Single_Use_Plastics.pdf.

5. Charles Moore with Cassandra Phillips, *Plastic Ocean: How a Sea Captain's Chance Discovery Launched a Determined Quest to Save the Oceans* (New York: Penguin Group, 2011).

6. Commoner, *Closing Circle*, 295.

7. World Economic Forum, *The New Plastics Economy*, January 2016, www3.weforum.org/docs/WEF_The_New_Plastics_Economy.pdf.

PART FIVE: TOWARD AN ECOLOGICAL CIVILIZATION

1. István Mészáros, *The Structural Crisis of Capital* (New York: Monthly Review Press, 2010), 35.

11. The MYTH OF "ENVIRONMENTAL CATASTROPHISM"

1. Fraser C. Lott, Nikolaos Christidis, and Peter A. Stott, "Can the 2011 East African Drought Be Attributed to Human-Induced Climate Change?," *Geophysical Research Letters* (March 2013): 1177–81.

2. UNDP, "'Rise of South' Transforming Global Power Balance, Says 2013 Human Development Report," March 14, 2013, http://www.undp.org/.

3. Tom Harris, "Climate Catastrophism Picking Up Again in the U.S. and Across the World," *Somewhat Reasonable Thoughts on Life and Liberty*, October 10, 2012, http://blog.heartland.org/2012/10/climate-catastrophism-picking-up-again-in-the-u-s-and-across-the-world/.

4. Pierre Gosselin, "The Climate Catastrophism Cult," *NoTricksZone*, February 12, 2011, http://notrickszone.com/.

5. Ray Evans, "The Chilling Costs of Climate Catastrophism," *Quadrant Online*, June 2008, http://quadrant.org.au/.

6. Franz Mauelshagen, "Climate Catastrophism: The History of the Future of Climate Change," in *Historical Disasters in Context: Science, Religion, and Politics*, ed. Andrea Janku, Gerrit Schenk, and Franz Mauelshagen (New York: Routledge, 2012), 276.

7. Alexander Cockburn, "Is Global Warming a Sin?," *CounterPunch*, April 28–30, 2007, http://www.counterpunch.org/.

8. Alexander Cockburn, "Who Are the Merchants of Fear?," *CounterPunch*, May 12–14, 2007, http://www.counterpunch.org/.

9. Alexander Cockburn, "I Am an Intellectual Blasphemer," *Spiked*, January 25, 2008, http://www.spiked-online.com/.

10. Leo Panitch and Colin Leys, preface to *Socialist Register 2007: Coming to Terms with Nature* (New York: Monthly Review Press, 2006), ix–x.

11. "Notes from the Editors," *Monthly Review* 58/10 (March 2007).

12. Sasha Lilley, David McNally, Eddie Yuen, and James Davis, *Catastrophism: The Apocalyptic Politics of Collapse and Rebirth* (Oakland, CA: PM Press, 2012).

13. Yuen's footnote cites an article that is identical to a news release issued the previous day by Texas A&M University; see "Increased Knowledge About Global Warming Leads to Apathy, Study Shows," *Science Daily*, March 28, 2008, https://www.sciencedaily.com/. The original paper, which Yuen does not cite, is P. M. Kellstedt, S. Zahran, and A. Vedlitz, "Personal Efficacy, the Information Environment, and Attitudes Towards Global Warming and Climate Change in the United State," *Risk Analysis* 28/1 (2008): 113–26.

14. Aaron M. McCright and Riley E. Dunlap, "The Politicization of Climate Change and Polarization in the American Public's Views of Global Warming, 2001–2010," *Sociological Quarterly* 52 (2011): 155–94.

15. Leiserowitz et al., "Global Warming's Six Americas," September 2012 (New Haven: Yale Project on Climate Change Communication, 2013), http://climatecommunication.yale.edu.

16. Robert J. Brulle, Jason Carmichael, and J. Craig Jenkins, "Shifting Public Opinion on Climate Change: An Empirical Assessment of Factors Influencing Concern over Climate Change in the U.S., 2002–2010," *Climatic Change* (September 2012): 169–88.

17. Joe Romm, "Apocalypse Not: The Oscars, The Media and the Myth of 'Constant Repetition of Doomsday Messages' on Climate," *Climate Progress*, February 24, 2013, https://thinkprogress.org/.

18. Neil T. Gavin, "Addressing Climate Change: A Media Perspective," *Environmental Politics* (September 2009): 765–80.

19. Two responses to David Noble are Derrick O'Keefe, "Denying Time and Place in the Global Warming Debate," *Climate & Capitalism*, June 7, 2007; and Justin Podur, "Global Warming Suspicions and Confusions," *ZNet*, May 11, 2007, https://zcomm.org/znet/.

20. Pablo Solón, "A Contribution to the Climate Space 2013: How to Overcome the Climate Crisis?," *Climate Space*, March 14, 2013, https://climatespace2013.wordpress.com/2013/03/14/a-contribution-to-the-climate-space-2013-how-to-overcome-the-climate-crisis/.

21. David Spratt, *Always Look on the Bright Side of Life: Bright-Siding Climate Advocacy and Its Consequences*, April 2012, http://climate-codered.org/.

22. Romm, "Apocalypse Not."

13. ECOSOCIALISM: A SOCIETY OF GOOD ANCESTORS

1.   Fidel Castro, "Tomorrow Will Be Too Late," in *The Global Fight for Climate Justice*, ed. Ian Angus (London: Resistance Books, 2009), 17–18.

2.   Constitution of the Republic of Cuba, Article 27, http://www.constitutionnet.org/files/Cuba%20Constitution.pdf.

3.   WWF, *2006 Living Planet Report*, 20, http://wwf.panda.org/.

4.   Richard Levins, "Living the 11th Thesis," in Richard Lewontin and Richard Levins, *Biology Under the Influence: Dialectical Essays on Ecology, Agriculture and Health* (New York: Monthly Review Press, 2007), 367.

5.   Armando Choy et al., *Our History Is Still Being Written* (New York: Pathfinder Press, 2006), 146.

6.   Karl Marx, *Capital,* vol. 3, in *MECW*, 35:763.

7.   Ian Angus, *Facing the Anthropocene*, 122.

8.   K. William Kapp, *Social Costs of Private Enterprise* (Cambridge, MA: Harvard University Press, 1950), 231, 62.

9.   Emily Matthews et al., *The Weight of Nations* (Washington, D.C.: World Resources Institute, 2000), xi.

10. Quoted in "Green Revolution with a Capital G Is Needed to Feed the World," UNEP Press Release, February 17, 2009, http://www.unep.org/.

11. Patrick McCully, "CDM Scams," *Climate & Capitalism*, May 22, 2008.

12. James Gustave Speth, *The Bridge at the Edge of the World: Capitalism, the Environment, and Crossing from Crisis to Stability* (New Haven: Yale University Press, 2009), 9 and 7.

13. Oswaldo Martinez, "We Are Facing Something More than a Mere Financial Crisis," translated by Richard Fidler, *Socialist Voice*, March 23, 2009, http://www.socialistvoice.ca/.

14. Levins, "Living the 11th Thesis," 370–71.

15. Levins, "How Cuba Is Going Ecological," in Lewontin and Levins, *Biology Under the Influence*, 344.

16. Evo Morales Ayma, "Evo Morales on Addressing Climate Change: 'Save the Planet from Capitalism,'" *Links International Journal of Socialist Renewal*, November 28, 2008, http://links.org.au/.

17. Barry Commoner, *The Closing Circle*, 295.

18. Ibid., 299–300.

Index